SpringerBriefs in Applied Sciences and Technology

SpringerBriefs present concise summaries of cutting-edge research and practical applications across a wide spectrum of fields. Featuring compact volumes of 50 to 125 pages, the series covers a range of content from professional to academic.

Typical publications can be:

- A timely report of state-of-the art methods
- An introduction to or a manual for the application of mathematical or computer techniques
- A bridge between new research results, as published in journal articles
- A snapshot of a hot or emerging topic
- An in-depth case study
- A presentation of core concepts that students must understand in order to make independent contributions

SpringerBriefs are characterized by fast, global electronic dissemination, standard publishing contracts, standardized manuscript preparation and formatting guidelines, and expedited production schedules.

On the one hand, **SpringerBriefs in Applied Sciences and Technology** are devoted to the publication of fundamentals and applications within the different classical engineering disciplines as well as in interdisciplinary fields that recently emerged between these areas. On the other hand, as the boundary separating fundamental research and applied technology is more and more dissolving, this series is particularly open to trans-disciplinary topics between fundamental science and engineering.

Indexed by EI-Compendex, SCOPUS and Springerlink.

More information about this series at https://link.springer.com/bookseries/8884

Ying Li

Motion Analysis of Soccer Ball

Dynamics Modeling, Optimization Design
and Virtual Simulation

 Springer

Ying Li
Houston, TX, USA

ISSN 2191-530X   ISSN 2191-5318   (electronic)
SpringerBriefs in Applied Sciences and Technology
ISBN 978-981-16-8651-1   ISBN 978-981-16-8652-8   (eBook)
https://doi.org/10.1007/978-981-16-8652-8

This Springer imprint is published by the registered company Springer Nature Singapore Pte Ltd.
The registered company address is: 152 Beach Road, #21-01/04 Gateway East, Singapore 189721,
Singapore

Acknowledgements

I am excited to announce my book titled Motion Analysis of Soccer Ball: Dynamics Modeling, Optimization Design and Virtual Simulation is published. It took me over one year to finish this. I credit this achievement in my hard work and self-discipline and even having supportive people. My teachers, editors, family, and friends offered their great assistance in constructing and writing this book. They deserve a word to be grateful.

I wish to express my deepest gratitude to my teacher, who provides an open course of engineering dynamics in a virtual classroom, for the valuable instruction. I appreciate his teaching throughout the whole class during the pandemic.

I would like to thank Dr. Mengchu Huang, Senior Editor of Applied Sciences Books in Springer, for his excellent editorial advice. With his suggestions, I have extended my two articles to the book. Without his advice, I might not decide to write this book.

I would like to give my special gratitude to my parents, two knowledgeable and educational persons. They contributed their wisdom to mentor me and guard me. No matter what happened, they always told me to be optimistic to face it. Without their spiritual encouragement, I might not have the strength to finish this book.

I thank Dr. Zhehui Liu, a scientist in the applied science field, for dedicated support and valuable discussions. He provided me with extraordinary information to help me to figure out how to build a complex model of soccer ball corner kick. He contributed his time to carefully review the Introduction Chapter and gave me some valuable suggestions for improving my book. I appreciate his advice and patience.

I want to thank Dr. Will Liu, an engineer, for providing all software I have used in the modeling and simulation of this book. Without those software, I would not be able to turn my ideas into virtual models.

I would like to thank Mr. Qi Li, a soccer coach, for his assistance in the preparation of Figs. 1.1–1.5 in the Introduction Chapter. He took all photos of physical prototypes that are used in this book for me to clearly indicate soccer field and structural components.

I would like to tell them that they almost did their best to support me to work on this book. Without their enthusiasm and encouragement, it would not have been possible for me to commit my research results to a book. Also, I gratefully acknowledge all contributors who related to this book! Their suggestions and the excellent advice provided me with ideas to dig every chapter with greater depth and finally bring the book to completion.

Subject Matter

The intelligent sports analysis of a soccer ball (also known as football, football ball, or association football ball) requires accurately simulating its motion and finding the best design parameters. Employing classic mechanics, the book establishes a fundamental framework for the soccer ball multi-body dynamics modeling, virtual prototype simulation, and optimization design. Five typical case studies have been addressed in the kinematics and dynamics simulations of soccer ball projectile motion, free kick, and corner kick in the virtual environment. The research on multi-dynamics models provides a useful method for engineers and scientists to investigate the spatial kinematics and dynamics performances of various balls, such as soccer ball, gulf ball, football, etc. The book is significant to guide undergraduate and graduate students from multi-disciplines to study system dynamics and optimization design. The book presents 3D virtual prototypes to illustrate the application for soccer players and trainers to predict the soccer ball trajectory.

Contents

1 Introduction 1
   1.1 Soccer Field and Structural Components 2
   1.2 Dynamics Modeling 6
   1.3 Model Validation 8
   1.4 Optimization Design 9
   1.5 Virtual Simulation 10
   1.6 Book Overview 10
   References 12

2 Dynamics Modeling and Numerical Validation of Soccer Ball
Projectile Motion 15
   2.1 Overview 15
   2.2 Dynamics Modeling of Soccer Ball Projectile Motion 15
       2.2.1 Initial Configuration 16
       2.2.2 Dynamics Modeling 18
       2.2.3 Solution 22
   2.3 Validation of Soccer Ball Dynamics Model 22
   References 29

3 Spatial Kinematics and Dynamics Simulation of Soccer Ball
Projectile Motion 31
   3.1 Overview 31
   3.2 Initial Configuration 31
   3.3 Dynamics Modeling 34
   3.4 Kinematics and Dynamics Simulation and Results Analysis 38
       3.4.1 Kinematics Simulation and Results Analysis 38
       3.4.2 Dynamics Simulation and Results Analysis 44
   References 48

4 Optimization Design of Soccer Ball Flight Trajectory 49
   4.1 Overview 49
   4.2 Optimization Modeling and Method Implementation 49

    4.3   A Case Study of a Soccer Ball Shooting at Target 52
          4.3.1   Initial Configuration and Dynamics Modeling 52
          4.3.2   Dynamic Simulation and Results Analysis 58
    4.4   Design Improvement Through Parameters Optimization 60
    References 65

5 Modeling and Simulation of Soccer Ball Free Kick 67
    5.1   Overview 67
    5.2   Initial Configuration 67
    5.3   Dynamics Modeling and Dynamic Simulation 70
          5.3.1   Dynamics Modeling 70
          5.3.2   Dynamic Simulation 75
    5.4   Simulation Results and Analysis 76
    References 81

6 Modeling and Simulation of Soccer Ball Corner Kick 83
    6.1   Overview 83
    6.2   Initial Configuration 83
    6.3   Dynamics Modeling 86
    6.4   Dynamic Simulation and Results Analysis 94
          6.4.1   Indirect Corner Kick 94
          6.4.2   Direct Corner Kick 98
    References 100

7 Contributions and Conclusions 101
    7.1   Dynamics Modeling 101
    7.2   Optimization Design 103
    7.3   Dynamic Simulation 103
    7.4   Future Works 104
    References 105

About the Author

Dr. Ying Li holds her B.E. and M.E. degrees in Mechanical Engineering from North University of China, Taiyuan, China, and the Ph.D. degree in Vehicle Engineering from University of Science and Technology Beijing, China. Her professional experience includes a faculty at North University of China; a Post-doctor Research Fellow at University of Missouri-Rolls, USA; a Post-doctor Research Fellow at University of Alberta, Canada; and a Principal Engineer at Caterpillar Global Mining, USA.

Dr. Li's research areas include engineering equipment dynamics modeling, virtual prototype simulation, structural strength analysis, mechanical-electric system co-simulation, multi-body dynamics full vehicle simulation. The results of her research initiatives include 3 patents, 10 books, 3 book chapters, over 60 peer-reviewed technical papers in international journals and conferences. She has reviewed dozens of articles for over ten international journals.

Dr. Li was one of the top ten winners of International Simulating Reality Contest, USA, 2015. She has been recognized with the Robert H. Quenon Endowed Research Fellowship Award by University of Missouri-Rolla, USA; the AERI/COURSE Research Fellowship Award by University of Alberta, Canada; Two Times of Distinguished Faculty Award, and Four Times of Outstanding Student Award by North University of China, China.

Chapter 1
Introduction

Soccer (also called as football) is the most popular sport in the world, which attracts millions of fans. Fans like to see the player shooting at goal. However, the score of a game is usually pretty low. The 0–0 game is frequently seen, meaning that it is not easy to achieve a successful outcome. Such final results make all fans disappointed. Soccer players all work very hard to keep in shape, and to improve their kicking skills. Outstanding soccer players can shoot at expected targets very successfully. Some of them result in goals. Therefore, shooting at a goal requires very high skills. However, most people don't know that intelligent sports analysis is a scientifically correct way of going about doing this.

The intelligent sports analysis requires exactly modeling the kinematics and dynamics of a soccer ball in a three-dimensional (3D) space. To address this problem, it is necessary to develop 3D dynamics models of soccer ball to predict the motion and capture the kinematics and dynamics performance. This book is targeted toward dynamics modeling, optimization design, and virtual simulation of soccer ball for efficient analysis of soccer ball motion.

The understanding of the soccer field and structural components is a basic step in the exploration of soccer ball kinematics and dynamics. Of great interest, to those who study the soccer ball are the aerodynamic properties, dynamic modeling and simulation, and trajectory analysis. This book is targeted toward the solution of dynamics modeling, optimization design, and virtual simulation problems in the quest for exactly predicting the flight trajectories of soccer balls by using new technologies. The introduction is involved in six aspects.

© The Author(s), under exclusive license to Springer Nature Singapore Pte Ltd. 2022
Y. Li, *Motion Analysis of Soccer Ball*,
SpringerBriefs in Applied Sciences and Technology,
https://doi.org/10.1007/978-981-16-8652-8_1

1.1 Soccer Field and Structural Components

Figures 1.1, 1.2, 1.3, 1.4 and 1.5 present that the actual soccer field, soccer ball, and facilities. All pictures were taken in the campus of North University of China, China, by Qi Li, who is a soccer coach and works at that university. He took these pictures for supporting me to write this book. I am thankful for his assistance. In Figs. 1.1, 1.2, 1.3, 1.4 and 1.5, it can find that in a soccer field, there are three major components, which are soccer field, soccer ball, and soccer players. Their functions, structures, conditions, and dimensions are introduced as below points.

The soccer fields are typically made of natural turf or artificial turf, and the fields are rectangular and have minimum and maximum dimensions as shown in Fig. 1.2a, b. The minimum overall dimensions are 90 m long and 45 m wide. The maximum overall dimensions are 120 m long and 90 m wide. For international matches: the minimum dimensions are the length of 100 m by the width of 64 m. The maximum dimensions are the length of 110 m by the width of 75 m.

Figure 1.2c is a two-dimensional soccer field. The middle of the field is divided lengthwise by the midfield line. In the center of the field, there is a big circle, which defenders must stay out of at the start of play. The four corner kick circles are marked on each corner of the field, indicating where a player must place the ball prior to a corner kick (Goldblatt 2008).

The goal area is the small box of 18.32 m wide by 5.5 m deep inside the penalty area as illustrated in Fig. 1.2c. The distance from each goalpost to the lines that goes parallel with the touch line should be 5.5 m. When the size of a standard goal is added it makes the distance between the lines 18.32 m (Goldblatt 2008).

Fig. 1.1 Soccer field, soccer ball, and soccer players (This picture was taken in the campus of North University of China, Taiyuan, China, by Qi Li)

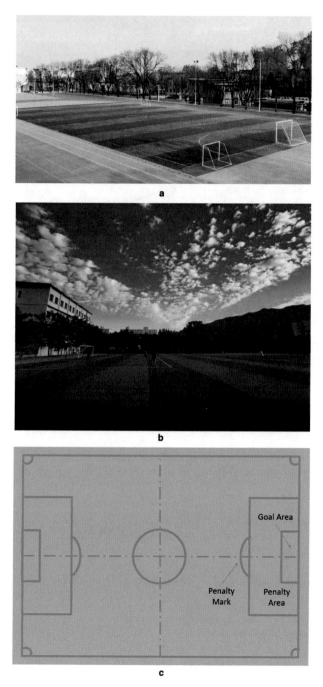

Fig. 1.2 a A small soccer field with facilities (This picture was taken in the campus of North University of China, by Qi Li). **b** A large soccer field with players (This picture was taken in the campus of North University of China, by Qi Li). **c** A two-dimensional soccer field

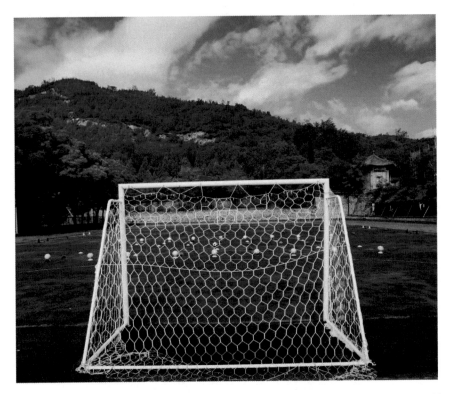

Fig. 1.3 A soccer goal (This picture was taken in the campus of North University of China, by Qi Li)

In Fig. 1.2c, it can be seen that a penalty area is the bigger box close to the goal that integrates the goal area and the penalty mark (D-shaped curve). It is 40.32 m long and 16.5 m wide. The lines that extend into the fields are 16.5 m and they have a 40.32 m distance from each other. The penalty mark is 11 m from the goal line and centered on the fields' goal line (Goldblatt 2008).

The soccer goal is constructed by one woodwork and one net as displayed in Fig. 1.3. The woodwork consists of two goalposts and one crossbar, which are made of wood, metal or other approved material. They must be square, rectangular, round or elliptical in shape and must not be dangerous to players. The standard soccer goal dimensions are 7.32 m wide and 2.44 m high. The crossbar length of 7.32 m is about 2.44 m above the ground. The post is 2.44 m high.

Figure 1.4a, b demonstrates some soccer balls, which are used in play. The soccer balls are made of both natural and man-made materials, such as plastic, rubber, synthetic cotton, and leather. The soccer balls are round (see Fig. 1.4b) and have different sizes for professional matches, children and adult games, international young men and young women matches, etc. The circumstance has to be minimum

a

b

Fig. 1.4 **a** Some soccer balls in the field (This picture was taken in the campus of North University of China, by Qi Li). **b** A soccer ball at the corner of the field (This picture was taken in the campus of North University of China, by Qi Li)

Fig. 1.5 Soccer players on the soccer field (This picture was taken in the campus of North University of China, by Qi Li)

0.69 m and maximum 0.71 m. The diameter has to be minimum 0.22 m and maximum 0.23 m. The weight has to be minimum 0.4 kg and maximum 0.45 kg. The internal pressure between 50.6 and 51.1 Pa.

Figure 1.5 shows some soccer players are playing a game. There are 11 soccer players on the soccer field per team during a soccer game. Each team has 10 field players and a goalkeeper.

1.2   Dynamics Modeling

How to shoot a soccer ball into a goal is also attracting scientists and engineers. In the past few years, the physics and mathematics of a soccer ball flight have been reported widely (Alam et al. 2010, 2011; Gupta and Panigrahi 2013; Asai et al. 2007, 2013; Hong et al. 2010, Dupeux et al. 2010, Hong and Asai 2011; NASA 2018; Goff and Carré 2009; Goff 2010; Bray and Kerwin 2003). In simple studies, a simple model used a simplified dynamics model with one free body (Sokolnikoff 1941; Meriam and Kraige 2002). The equation of motion in the simple models includes only the gravity applied on the ball. This model may apply to the penalty kick and goalkeeper kick. The ball traveling trajectory is similar to the traveling trajectory of a flying bullet.

However, the simple model cannot be applied to the direct free kick or corner kick. The ball traveling trajectory of a successful direct free kick likes a banana so that the ball flies over the player wall to the goal. A model predicting a banana free kick must include both the effects of gravity, aerodynamic characteristics (such as the aerodynamic drag and drag moment), and Magnus. The aerodynamics of soccer balls is important in emphasizing that a ball flight derives from not only gravity but also aerodynamic drag or Magnus effects. Magnus effects can produce non-symmetrical airflow and a differential force that leads to spin the ball in flight.

The studies involved the theoretical or experimental investigations of the direct free kick in soccer ball (NASA 2018; Goff and Carré 2009; Bray and Kerwin 2003). The aims were to develop the mathematical models of the ball's flight incorporating aerodynamic lift and drag forces to explore flight trajectories. Trajectories derived from the models have been compared with those obtained from detailed video analysis of experimental kicks. Representative values for the drag and lift coefficients have been obtained. These values, used with a simple model of a defensive wall, have enabled free kicks to be simulated under realistic conditions, typical of match-play. The results reveal how carefully attackers must engage the dynamics of a successful kick.

Javorova and Ivanov (2018) and Zhu et al. (2017) have constructed a training system to simulating free kicks or corner kick for training players. Employing the dynamics principle, Bernoulli principle, and Magnus effect equation, they developed the mathematical models of soccer ball flights. The effects of the initial velocity, kick force, and kick position on the traveling trajectory were investigated. Additionally, suggestions are given on how to shoot at a goal, avoiding the defensive walls and goalkeeper through optimizing the parameters of the kicking point, the kicking force, and the kick angle.

The above research is all based on one free-body dynamics. The traveling trajectory can be predicted. They can only apply to free kick or corner kick when the ball first hits a target. However, when the ball bounces into the goal, the traveling trajectory cannot be predicted. The influence of impact forces on a soccer ball's flight has received little attention. Levendusky et al. (1988) has started of impact characteristics of soccer balls, commented on the difficulty of dropping them with repeatable accuracy onto a force plate from a height of 18.1 m. The variability was attributed to aerodynamic drag forces and Magnus effects, although no quantitative information was given to support the statement. This lack of committed research on spin effects in soccer is surprising, since the technique is widely used by players in many aspects of the game, especially when trying to beat the defensive wall with a direct free kick (Bray and Kerwin 2003).

Recently, the development of multi-body dynamics model has made it is possible to apply the model to free kick when the ball hits the field or goal (Li and Li 2019; Li et al. 2020). Li and Li have created a two-body dynamics model of soccer ball-field. Then, Li et al. have extended the two-body model to a three-body dynamics model of the field-goal-ball. The contact force among the ball, goal, and field has been integrated into the model. The model can be used to predict the bouncing trajectory.

These studies have the application of the work in soccer free kick. But the application in corner kick has received a limited discussion.

All above works employed Newton–Euler equations to build the dynamics equations of soccer ball. Newton–Euler method is of importance for real-time simulation and parameter estimation. Also, it provides detailed information on all positions, which will be useful in possible motion and/or force analysis.

Further studies have toward applying the milt-body dynamics model to indirect free kick and corner kick for simulating full kinematics and dynamics performance. The complex traveling trajectory can be predicted when the ball is touched by a player and bounces into the goal. The model can allow randomly applying a kick force or initial velocity on the ball in 3D space.

1.3 Model Validation

The model examination and validation can correct dynamics model for obtaining accurate simulation results. Previous research has conducted various laboratory tests to verify the dynamic models of soccer ball. Asai et al. (2007, 2013). conducted a steady-state analysis of the newest soccer ball and conventional soccer balls through a wind tunnel experiment and clarified the drag coefficient and critical Reynolds number. A simple 2D flight trajectory simulation was conducted based on the drag coefficient, and the effects of the drag characteristics on the flight distance and flight trajectory were examined. The two were shown to have a high degree of correlation.

Goff and Carré (2009, 2010, 2012) performed experiments in which a soccer ball was launched from a machine while two cameras recorded portions of its trajectory. The no-spin trajectories were obtained from range measurements during the ball's flight. The experiment methods of measuring trajectories are available to educators and students to validate the estimated results from a simple computational approach.

These methods are very efficient, but their processes are costly and time-consuming for achieving statistically significant basis for evaluation and subsequent decisions. With the development of modern virtual techniques, computer modeling and simulation have been applied in the development, design, and analysis of soccer ball motion (Javorova and Ivanov 2018; Zhu et al. 2017; Li 2019; Li et al. 2020). Computer design processes in virtual simulation environments enable the testing of soccer ball trajectory prior to final field-testing.

A substitute method can be numerical validation (Maczynski and Szczotka 2002; Li and Liu 2014) employing computer simulation with virtual models, such as develop two or more computer models using different methods, and do same testing, then analyze and correct the result of the data. The important concern is the model should be as simple as possible, but it must sufficiently correct to reflect the dynamic features.

Li and Li (2020) have used this method to validate a soccer dynamics model through a comparison of their simulation result and the computational result (Hroncová and Grieš 2014). An example is presented for demonstrating the model application as well as kinematics and dynamics analysis. The results show that dynamic analysis can reflect the dynamic characteristics of soccer ball, which is missed and not reflected from static analysis. This research provides a foundation for the validation of soccer dynamics model. However, the numerical model validation of soccer ball is still in the preliminary stage and more efforts are needed.

1.4 Optimization Design

How to plan an expected trajectory of soccer ball without practicing on a field is a significant topic in soccer research. An optimization method can help to carry out the trajectory planning to improve the design parameters and optimize the flight trajectory. The optimization method has been applied to the engineering design since the 60–70 s (Powell 1964; Johnson 1971; Fox 1972). Gradually, the method is widely used in mechanical engineering for optimizing mechanism design (ANSYS 2013). Then, the optimization method is extended to sports analysis.

Some researchers have proposed optimal models to capture the behavior of a ball being impacted. Christenson (2018) examines the physics behind the optimal football pressure as a function of ball speed, touch, and force of kick, considering the football modeled as two lumped masses and the kicking foot modeled as a spring mass system with spring interface with the ball. The model is able to capture the dynamics of the foot impacting a ball and associated dynamics of the ball itself. Simulations were conducted and the optimal football pressure is proposed as a function of footballer's ability by considering a multi-objective DIRECT optimization. In such a manner, the optimal football pressure can be identified for players of distinct ability.

A basketball dynamics model was developed for achieving trajectory optimization (MSC Software 2018). The case study has focused on demonstrating the optimization operation. The model itself is overly simplified, by neglecting either the aerodynamic characteristics or the initial configuration. However, these factors have been shown to be important in the dynamics simulation of flight objects to capture their three-dimensional (3D) nature.

Recently, Li et al. (2020) have extended the two-body model of field-ball to a three-body dynamics model of the field-goal-ball. An optimization model is integrated into the model to predict the traveling trajectory. The ball's original trajectory is simulated, and then the optimization operation is applied to find an optimized path. A case study is presented to illustrate the method application. This topic has well been expanded into a complete chapter in this book.

1.5   Virtual Simulation

There were a lot of studies about the flight trajectories of soccer ball. The results have been output as a series of displacement–time curves. For soccer players and trainers, it may be hard to imagine the position of a ball in 3D space. It is hard to find the distribution of dynamic forces. Accurate representations of flight trajectory and simultaneous position in soccer field require a virtual reality environment for comprehensive dynamic simulation. The virtual prototype technology solves the problem of visualization to check design based on computer science and technology (Zorriassatine et al. 2003).

In recent years, the virtual reality techniques have been applied in the development, design, and analysis of flight objects' motion (Javorova and Ivanov 2018; MSC Software 2018; Hroncová and Grieš 2014; Zhu et al. 2017; Li 2019; Li et al. 2020). The motion of the object can be visualized by plotting successive positions on graphic displays. Typical virtual simulation represents the design in terms of wireframe and surface. Some of them are only a geometric representation of the object for observation. They do not have enough features to allow simulating, testing, and analyzing the model against the requirement. These representations have little significance in design examination and investigation of kinematics and dynamics performance.

The rapid increase in computing power and speed enables the development of a fully functional virtual prototype just as to build a physical prototype. Some studies (MSC Software 2018; Hroncová and Grieš 2014; Li 2019; Li et al. 2020) have integrated 3D solid modeling, dynamics modeling, and dynamic simulation into a virtual environment. In their studies, objects are created as virtual prototypes. The motion of the object is established as the time-dependent ordinary differential equations (ODEs). The dynamic simulations focus on the visualization of the object traveling trajectory and the plotting curves of the time-varying displacement or force. In case studies (Li and Li 2019; Li et al. 2020), the animation provides a function to display the instantaneous positions of the soccer ball with curving and bending postures. Also, the instantaneous force applied to the ball is visualized and plotted with the simulating time, which is useful to analyze the dynamic performance.

Therefore, the virtual prototype technology is very useful to visualize soccer ball motion and dynamic force distribution in 3D space. It is enabled successive ball positions to be accurately determined. It builds and simulates realistic full-motion behavior models with capabilities for quick analysis of multiple design variations toward an optimal design.

1.6   Book Overview

This book introduces dynamics modeling, optimization design, and virtual simulation of soccer ball. It combines dynamics, mathematics, and sports engineering from a multi-disciplinary design optimization to lay a foundation for the full dynamic simulation of soccer ball motion. Chapters 2–6 discusses five subjects: model validation, projectile motion, trajectory optimization, free kick, and corner kick. Each of these highlights different aspects of soccer ball kinematics and dynamics. Some typical examples are given to indicate the practical applications. All expectations are involved in six aspects.

Chapter 2: Dynamics Modeling and Numerical Validation of Soccer Ball Projectile Motion. The main steps are introduced to build a basic dynamics model of soccer ball. A numerical validation method is suggested to validate the dynamics models of soccer ball. The virtual prototype technology is applied for the solution of model validation problems in the quest for using a virtual simulation instead of a physical simulation. A case study illustrates the dynamics model is highly reliable in predicting the flight trajectory of a soccer ball.

Chapter 3: Spatial Kinematics and Dynamics Simulation of Soccer Ball Projectile Motion. A spatial two-body dynamics model of a soccer field-ball has been developed. The model includes gravity, aerodynamic drag, and contact force. The instantaneous position and force applied to the ball are visualized through virtual kinematics and dynamics simulation. All works form a complete procedure to illustrate the application of dynamic simulation on soccer ball projectile motion analysis.

Chapter 4: Optimization Design of Soccer Ball Flight Trajectory. An optimization method has been proposed to predict a soccer ball target. The parameterization technology has been integrated into the optimization design for automatically capturing the expected flight trajectory. The expected target is expressed as a function of all design parameters. A case study goes through multi-body dynamics modeling, dynamic simulation, and optimal objective modeling. The study indicates the application of the optimization design method on the automatic dynamic analysis of soccer ball shooting at goal.

Chapter 5: Modeling and Simulation of Soccer Ball Free kick. For the study of free kick, the spatial two-body dynamics model is extended to the three-body dynamics model of a soccer field-goal-ball. The model includes gravity, aerodynamic drag, Magnus force and contact force. The typical cases of free kick are simulated, and animation displays the successive ball positions with curving, bending, and spinning postures. The result analysis is addressed.

Chapter 6: Modeling and Simulation of Soccer Ball Corner Kick. An innovative multi-body dynamics modeling of the soccer field-goal-ball-player has been proposed by integrating Newton's section law and Hooke's law. The model has mostly embedded the gravity, aerodynamic drag, drag moment, Magnus force, impulsive force, and contact force. The model allows randomly applying a kicking force or initial velocity on the ball in 3D space. Case studies of indirect and direct corner kick are focused on simulating reality and analyzing results.

Chapter 7: Contributions and Conclusions. The exact contributions of the works in this book are discussed and compared to recently published works. Some important conclusions are summarized, and future works are suggested. The book provides valuable guidance for dynamics modeling, model validation, multi-disciplinary design optimization, and virtual prototype simulation of soccer ball and further improvement.

References

Asai T, Seo K, Kobayashi O, Sakashita R (2007) Fundamental aerodynamics of the soccer ball. Sports Eng 10:101–109

Asai T, Seo K (2013) Aerodynamic drag of modern soccer balls. Springerplus 2:171

Alam F, Chowdhury H, Moria H (2010) A comparative study of football aerodynamics. J Procedia Eng. 2:2443–2448

Alam F, Ho H, Chowdhury H, Subic A (2011) Aerodynamics of Baseball. J Procedia Eng 13:207–212

Ansys Software (2013) Design optimization. University of Alberta, Canada, Tutorial Book

Bray K, Kerwin DG (2003) Modelling the flight of a soccer ball in a direct free kick. J Sports Sci 21:75–85

Christenson A, Cao P, Tang J (2018) Optimal football pressure as a function of a footballer's physical abilities. Presented at the 12th conference of the international sports engineering association, Brisbane, Queensland, Australia, 26–29 March

Dupeux D, Le Goff A, Quéré D, Clanet C (2010) The spinning ball spiral. New J Phys 12:093004

Fox RL (1972) Optimization methods for engineering design. Addison-Wesley Publisher

Goldblatt D (2008) A global history of football. FootballHistory.org

Goff JE, Carré MJ (2009) Trajectory analysis of a soccer ball. Am J Phys 77:1020–1027

Goff JE, Carré MJ (2010) Soccer ball lift coefficients via trajectory analysis. Eur J Phys 31:775–784

Goff JE (2010) Power and spin in the beautiful game. Phys. Today, 62–63

Goff JE, Carré MJ (2012) Investigations into soccer aerodynamics via trajectory analysis and dust experiments. Procedia Eng. 34:158–163

Gupta G, Panigrahi PK (2013) Curve kick aerodynamics of a soccer ball. Proceeding of the fortieth national conference on fluid mechanics and fluid power, Himachal Pradesh

Hroncová D, Grieš M (2014) Trajectories of projectiles launched at different elevation angles and modify design variable in MSC Adams/View. Appl. Mech. Mater. **611**, 198–207. ISSN: 1662–7482

Hong S, Chung C, Nakayama M, Asai T (2010) Unsteady aerodynamic force on a knuckleball in soccer. Procedia Eng 2:2455–2460

Hong S, Asai T (2011) Aerodynamics of knuckling effect shot using kick-robot. Int J Appl Sports Sci 23(2):406–420

Johnson RC (1971) Mechanical design synthesis with optimization application. New York, USA

Javorova J, Ivanov A (2018) Study of soccer ball flight trajectory. MATEC web of conferences 145:01002

Levendusky TA, Armstrong CW, Eck JS, Spryropoulous P, Jeziorowski J, Kugler L (1988) Impact characteristics of two types of soccer balls. In: Reilly T, Lees A, Davids K, Murphy WJ (eds) Science and Football. E & FN Spon, London, pp 385–393

Li Y, Li Q (2019) Soccer ball spatial kinematics and dynamics simulation for efficient sports analysis. Asian J Adv Res Reports 7(4):1–18 ISSN: 2582–3248

Li Y, Liu WY (2014) Comparison of advanced dragline dynamics models for efficient engineering analysis. Int J Adv Manuf Technol 72(5–8):757–764

Li Y, Jx M, Li Q (2020) Predicting soccer ball target through dynamic simulation. J Eng Res Reports 12(4):6–18

Maczynski A, Szczotka M (2002) Comparison of models for dynamic analysis of a mobile telescope crane. J Theor Appl Mech 40(4):1051–1074

Meriam JL, Kraige LG (2002) Engineering mechanics, dynamics, 5th edition. Wiley, Incorporated, New York, USA

MSC Software (2018) Automatic dynamic analysis of mechanical systems. University of Michigan, USA, Technic Manual

NASA (2018) Drag on a soccer ball. National Aeronautics and Space Administration Technic Paper, www.nasa.gov

Powell MJD (1964) An efficient method for finding the minimum of a function of several variables without calculating derivatives. Comput J

Sokolnikoff IS, Sokolnikoff ES (1941) Higher mathematics for engineers and physicists. McGraw Hill Book Company Inc., New York, USA

Zhu ZQ, Chen B, Qiu SH, Wang RX (2017) Simulation and modeling of free kicks in football game and analysis on assisted training. Asia Sim 2017, Part I, CCIS751, pp 413–427

Zorriassatine F, Wykes C, Parkin R, Gindy N (2003) A survey of virtual prototyping techniques for mechanical product development. Proc IME B J Eng Manufact 217(2):513–518

Chapter 2
Dynamics Modeling and Numerical Validation of Soccer Ball Projectile Motion

2.1 Overview

The kinematics and dynamics modeling are an important factor to realize intelligent analysis of a soccer ball motion in a virtual environment. The problem of how to accurately create a three-dimensional (3D) soccer model is addressed in this Chapter. The approach of dynamics modeling is proposed towards the precisely predicting the flight trajectory of a soccer ball and capturing its kinematics and dynamics performance. The model is governed by the classic mechanics and formulated as time-dependent ordinary differential equations (ODEs). The numerical method of soccer modeling validation is suggested and demonstrated by employing a simple case. The projectile motion of a soccer ball is presented. The model is validated by the comparison of the simulated result and computed result. This investigation has laid a foundation for the dynamics modeling and accurate simulation of a soccer ball motion towards the planning, evaluating, and optimizing its trajectory.

2.2 Dynamics Modeling of Soccer Ball Projectile Motion

A dynamics model is built for the dynamic simulation of a soccer ball projectile motion. The model is created by ignoring the ball flying spin, Magnus effect, drag moment, gyroscopic moment, and acceleration of Coriolis. The model consists of several sets of mathematical equations, which are the equation of football impulsive force, the equations of initial velocity, the equations of aerodynamic drag applied on the ball, and the equations of ball-field contact at the landing point. The modeling details are given in the following aspects.

© The Author(s), under exclusive license to Springer Nature Singapore Pte Ltd. 2022
Y. Li, *Motion Analysis of Soccer Ball*,
SpringerBriefs in Applied Sciences and Technology,
https://doi.org/10.1007/978-981-16-8652-8_2

2.2.1 Initial Configuration

Figure 2.1 shows a soccer ball projectile motion through the air on a soccer field. A Cartesian coordinate system o(x, y, z) is attached to the field at the ball center of mass in initial position, where x–z plane is attached on the soccer field with x-axis along the length direction and z-axis along the width direction, as well as y-axis is perpendicular to the x–z plane.

Figure 2.2 shows a free-body diagram of a ball with the initial configuration of a kicking force and two velocity vectors. A ball with mass m is moving at velocity $\overrightarrow{v}_1$ on a given direction and a player kicks the ball at its center with force $\overrightarrow{P}$. When the ball leaves the player's foot, the ball travels with velocity $\overrightarrow{v}_2$. If the foot and ball are in contact with time Δt, the relationship between the force vector $\overrightarrow{P}$ and velocity vectors $\overrightarrow{v}_1$ and $\overrightarrow{v}_2$ can be given by the principle of linear impulse and momentum (Meriam and Kraige 2002) as illustrated in Eq. 2.1.

$$\overrightarrow{P}\,\Delta t = m\,\overrightarrow{v}_2 - m\,\overrightarrow{v}_1 \tag{2.1}$$

$$\overrightarrow{P} = \frac{m\,\overrightarrow{v}_2 - m\,\overrightarrow{v}_1}{\Delta t} \tag{2.1a}$$

$$\overrightarrow{v}_2 = \frac{\overrightarrow{P}\,\Delta t + m\,\overrightarrow{v}_1}{m} \tag{2.1b}$$

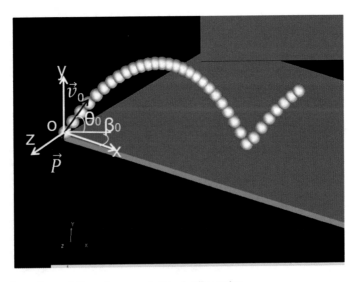

Fig. 2.1 Dynamics modeling of a soccer ball projectile motion

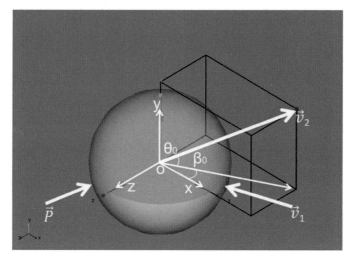

Fig. 2.2 Free-body diagram of a ball with a kicking force and velocity vectors

Usually, two problems are solved (i) given velocities $\vec{v}_1$ and $\vec{v}_2$ to find the magnitude and direction of the average impulsive force $\vec{P}$ (see Eq. 2.1a); and (ii) reversely, given impulsive force $\vec{P}$ and $\vec{v}_1$ to find the magnitude and direction of $\vec{v}_2$ (see Eq. 2.1b).

Here, the study focuses on projectile motion. As shown in Fig. 2.3, $\vec{v}_2$ is set as initial velocity expressed as $\vec{v}_0$ and its direction is defined as initial projectile angle θ_0, and initial orientation angle β_0. Initial velocity vector $\vec{v}_0$ can be broken down into three scalar components $v_{x0}, v_{y0},$ and v_{z0} in x–y-z directions as illustrated in Eqs. 2.2a–2.2c.

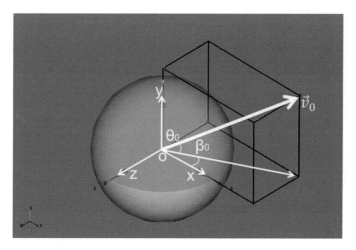

Fig. 2.3 Free-body diagram of a ball with an initial velocity

$$v_{x0} = v_0 \cos\theta_0 \cos\beta_0 \qquad\qquad (2.2a)$$

$$v_{y0} = v_0 \sin\theta_0 \qquad\qquad (2.2b)$$

$$v_{z0} = -v_0 \cos\theta_0 \sin\beta_0 \qquad\qquad (2.2c)$$

2.2.2 Dynamics Modeling

Figure 2.4 shows the soccer ball projectile motion and a free-body diagram of a soccer ball in an instantaneous position. The model is created by assuming that the air around the ball is non-homogenous, and ignoring the ball flying spin, Magnus effect, drag moment, gyroscopic moment, and acceleration of Coriolis. The forces applied on the ball include gravity $\vec{G}$, aerodynamic drag $\vec{F}_d$ and contact force $\vec{F}_n$. The equations of three forces are established as the following form.

Gravity: gravity $\vec{G}$ acts to the ball in $-y$ direction and is calculated by Eq. 2.3.

$$\vec{G} = -mg \qquad\qquad (2.3)$$

where g is the gravitational acceleration. Goff and Carré (2009) assume that the trajectory of a soccer ball is close enough to the earth's surface so that the gravitational acceleration can be considered as constant $g = 9.8$ m/s^2.

Aerodynamic Drag: when the ball flies through the air, the air resists the ball motion with aerodynamic drag $\vec{F}_d$. The aerodynamic drag can be described as force, which

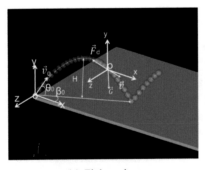

(a) Fight trajectory

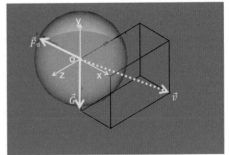

(b) Free-body diagram

Fig. 2.4 Soccer ball flight trajectory and a free-body diagram at an instantaneous position. **a** Fight trajectory **b** Free-body diagram

acts against the flight direction of the soccer ball. It depends on the density of the air ρ, the cross-sectional area of the ball A, the velocity of the ball moving $\vec{v}$ and the drag coefficient C_d. Then, the aerodynamic drag is established as Eqs. 2.4 and 2.5 (Bray and Kerwin 2003; Asai et al. 2007, 2013; Goff and Carré 2009).

$$\vec{F}_d = -\frac{1}{2}\rho C_d A \vec{v} \lfloor \vec{v} \rfloor = -K_d \vec{v} |\vec{v}| \tag{2.4}$$

If d denotes the diameter of the ball, then,

$$K_d = \frac{1}{8}\rho C_d \pi d^2 \tag{2.5}$$

There are two important components of the aerodynamic drag that are relevant. One is the density of the air ρ, which can be calculated by the air pressure divided by the specific gas constant for dry air and the temperature in Kelvin (Atkins and De Paula 2008). Another one is the drag coefficient, which depends on the boundary conditions, such as the surface roughness of a ball, as well as the laminar flow or turbulence of the atmospheric layer. It is an important parameter to describe the behavior of a soccer ball during flying through the air and can be calculated based on the investigations of Asai et al (2007) and Smith et al. (1999). The drag coefficients of a soccer ball are suggested between 0.1 and 0.5 (Smith et al. 1999). In this book, $C_d = 0.2$ is selected for all case studies.

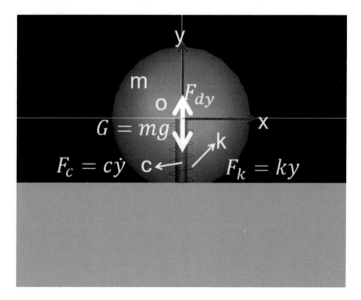

Fig. 2.5 Ball-field contact model with a ball mass representation of a ball hitting a field with spring stiffness and damping dashpot

Figure 2.5 shows a free-body diagram of the ball in an instantaneous position. It notes that aerodynamic drag $\overrightarrow{F}_d$ aligns and opposites to the direction of velocity $\overrightarrow{v}$. K_d is related to the density of the air ρ, drag coefficient C_d and diameter of the ball d. If the air condition is non-homogenous, K_d will be broken down into three components K_{dx}, K_{dy}, and K_{dz} along with x–y–z directions, respectively. Vector Eq. 2.4 can be represented as three sets of scalar Eqs. 2.6a–2.6c. Therefore, aerodynamic drag $\overrightarrow{F}_d$ can be separated into three component forces F_{dx}, $F_{dy,}$ and F_{dz}, which are applied to the ball in x–y–z directions, respectively.

$$F_{dx} = -K_{dx}\dot{x}|\dot{x}| \tag{2.6a}$$

$$F_{dy} = -K_{dy}\dot{y}|\dot{y}| \tag{2.6b}$$

$$F_{dz} = -K_{dz}\dot{z}|\dot{z}| \tag{2.6c}$$

The dynamics equation of the ball is governed by Newton–Euler Equations (Synge and Griffith 1959).

$$\sum \overrightarrow{N} = m\overrightarrow{a} = \overrightarrow{G} + \overrightarrow{F}_d \tag{2.7}$$

where $\sum \overrightarrow{N}$ is the vector summation of the general external forces; and $\overrightarrow{a}$ is the acceleration vector of the ball. Equation 2.7 can be expressed as three ODEs by separating it into the following three sets of scalar Eqs. 2.8a–2.8c.

$$m\ddot{x} = -K_{dx}\dot{x}|\dot{x}| \tag{2.8a}$$

$$m\ddot{y} = -mg - K_{dy}\dot{y}|\dot{y}| \tag{2.8b}$$

$$m\ddot{z} = -K_{dz}\dot{z}|\dot{z}| \tag{2.8c}$$

Ball-field Contact Force: when the ball lands on the field, the ball impacts the field and finishes one cycle of the projectile motion. Figure 2.5 shows a diagram of the final configuration of ball-field contact. A simplified lumped mass dynamic model is proposed to capture the dynamics of the ball hitting the field. The ball-field contact model is created as a mass-spring-dashpot system with one degree of freedom to generate the interaction between the ball and field. The mass m is equal to the mass of the ball, the spring represents the ball's elastic properties and the dashpot expresses the ball's damping characteristics. There are three types of forces applied to the system, internal forces, external forces, and inertia force.

The internal forces are caused by the spring and damping, which act at the contact point in the y-direction, which is normal to the ball-field contact patch. This normal reaction force F_n has two components, a spring force $F_k = ky$, which is proportional to contact stiffness k and deflection y, and opposite to the relative deflation; and a damping force $F_c = c\dot{y}$, which are proportional to the contact damping coefficient c and the rate of deflation, and opposite to the direction of relative motion. Therefore,

$$F_n = F_k + F_c = c\dot{y} + ky \qquad (2.9)$$

Note the contact stiffness and contact damping coefficient cannot find in material property handbooks. The data given in this book are calculated results. The natural frequency of the ball ω_n can be written as function of the mass and stiffness as

$$\omega_n = \sqrt{\frac{k_1}{m}} \qquad (2.10)$$

then, the ball stiffness k_1 is

$$k_1 = \omega_n^2 m \qquad (2.11)$$

The field stiffness k_2 can be found in material property handbooks. The contact stiffness k is defined as

$$k = \frac{1}{\frac{1}{k_1} + \frac{1}{k_2}} \qquad (2.12)$$

The damping coefficient of the ball c_1 can be written as function of the mass and natural frequency as

$$c_1 = 2zm\omega_n \qquad (2.13)$$

where z is the damping ratio for the ball. Christenson et al. suggest z = 12% (Christenson et al. 2018). The damping coefficient of the field c_2 can be found in material property handbooks. The contact damping coefficient c is defined as

$$c = \frac{1}{\frac{1}{c_1} + \frac{1}{c_2}} \qquad (2.14)$$

In the next four chapters, the contact stiffness k = 97.2 N.s/m and the contact damping coefficient c = 3.8×10^5 N/m will be used for all case studies.

The external forces come from gravity G and aerodynamic drag in y-direction F_{dy}.

The inertia force is related to the mass of the ball and the acceleration of deflation, which is also applied to the ball in the y-direction. The equations of motion for the ball-field contact are given as Eqs. 2.15 and 2.16.

$$m\ddot{y} = \sum N_y = -G - F_{dy} + e(F_c + F_k) \tag{2.15}$$

$$m\ddot{y} = -mg - K_{dy}\dot{y}|\dot{y}| + e(c\dot{y} + ky) \tag{2.16}$$

The general dynamics equations are given as Eqs. 2.17a–2.17c, in x–y-z directions, respectively.

$$m\ddot{x} = -K_{dx}\dot{x}|\dot{x}| \tag{2.17a}$$

$$m\ddot{y} = -mg - K_{dy}\dot{y}|\dot{y}| + e(c\dot{y} + ky) \tag{2.17b}$$

$$m\ddot{z} = -K_{dz}\dot{z}|\dot{z}| \tag{2.17c}$$

where, e = 0, when y > = d/2, there is no penetration between the ball and field and the contact force is zero; and e = 1, when y < d/2, there is penetration between the ball and field and the contact force is larger than zero.

2.2.3 Solution

Figure 2.6 shows a flow chart of dynamics modeling and solution process of soccer ball motion. Integrating the initial conditions (Eqs. 2.2a–2.2c) into ordinary differential Eqs. 2.17a–c builds a dynamics simulation system for capturing kinematics and dynamics performance of a soccer ball projectile motion. The equations of the impulsive force, projectile motion, and ball-field contact force can be solved using a computer program. The numerical results give the time-varying aerodynamic drag, projectile motion (displacements, velocities, accelerations), and contact forces of the ball-field. The animation can visualize the projectile trajectory of the soccer ball as demonstrated in Figs. 2.7 and 2.8 Dynamic simulation diagram of the soccer ball projectile motion (Li and Li 2019).

The solution for these ODEs supports the sports analysis of soccer ball. The dynamics models can help players to find the kicking force, kicking position, and kicking angle to predict an expected trajectory. The dynamic simulation can be used for the motion and dynamic force analysis for training players.

2.3 Validation of Soccer Ball Dynamics Model

The validation is necessary to verify the proposed dynamics model of soccer ball. The laboratory test of soccer ball motion is the best way to validate the dynamics model. But it is very costly and time-consuming for achieving statistically significant basis

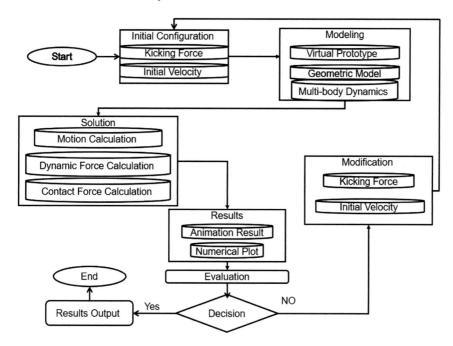

Fig. 2.6 Flowchart of dynamics modeling and solution process of soccer ball motion

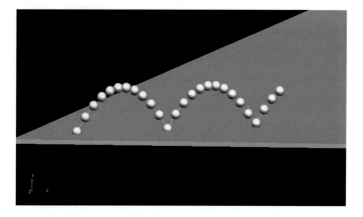

Fig. 2.7 Animation displaying the projectile trajectory of a soccer ball (Li and Li 2019)

for evaluation and subsequent decisions. A substitute method is numerical dynamic analysis (Maczynski and Szczotka 2002; Li and Liu 2014) employing computer simulation with mathematical models, such as develop two or more computer mathematical models using different methods, and do same testing, then analyze and correct the result of the data. The important concern is the model should be as simple as possible, but it must sufficiently correct to reflect the dynamic features.

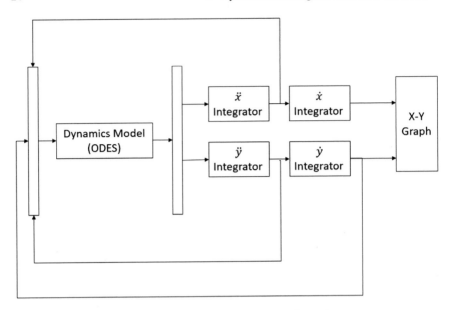

Fig. 2.8 Dynamic simulation diagram of the soccer ball projectile motion

Ballistics is the study of objects flying through the air, one that has received considerable attention owing to its military applications (Hroncová and Grieš 2014). Li and Li (2019) have extended this simplest possible theoretical description of a soccer ball in flight, that of a sphere of mass m flying through the air, in which there is the influence of gravity and aerodynamic drag on the motion, a classic aerodynamics problem. In this Chapter, the dynamics model of a soccer ball is validated by comparing the results of the dynamic simulation (Li and Li 2019) against those of a mathematical calculation (Hroncová and Grieš 2014). The comparison of displacement–time curves is presented in a two-dimensional (2D) Cartesian coordinate system o(x, y). The details are discussed in below.

Dynamic Simulation: a soccer ball projectile motion is simulated using the dynamics model created in this Chapter. The model is created by assuming air condition is homogenous. Equations 2.11 a–b are applied to the 2D model in a coordinate system o(x, y) and is combined to form a matrix as expressed in Eq. 2.18. There are two equations and two unknowns. Figure 2.8 shows the dynamic simulation diagram of this simple system. There is no input system. The initial conditions will determine its response.

$$\begin{bmatrix} m & 0 \\ 0 & m \end{bmatrix} \begin{bmatrix} \ddot{x} \\ \ddot{y} \end{bmatrix} = \begin{bmatrix} -K_d \dot{x}|\dot{x}| \\ -mg - K_d \dot{y}|\dot{y}| + e(c\dot{y} + ky) \end{bmatrix} \tag{2.18}$$

Table 2.1 Geometric parameters, physical property, and initial parameters

Geometric parameters		Physical property	Initial parameters		
Ball diameter	Ball mass	Coefficient	Initial velocity	Initial projectile angle	Initial orientation angle
d (m)	m (kg)	K_d (kg/m)	v_0 (m/s)	θ_0 (°)	β_0 (°)
0.2286	0.43	0.00002	46.5	45	0

For comparing purpose, the parameters, such as initial conditions and simulation time are selected as same as those used in a case study reported by Hroncová and Grieš. The general information: geometric parameters, physical property, and initial parameters, is listed in Table 2.1. They are the ball diameter d = 0.2286 m, the mass of the ball m = 0.43 kg, coefficient K_d = 0.00002 kg/m, initial velocity v_0 = 46.5 m/s, initial projectile angle θ_0 = 45°, and initial orientation angle β_0 = 0°.

Figure 2.9a–c shows the dynamic simulation of the soccer ball versus (vs) time 7 s (s). The soccer ball is launched at an initial point o(0, 0) and landed on the field at point (226, 0). Figure 2.9a plots the x–y displacement of the ball in coordinate system o(x, y). It notes that the ball traveling trajectory exhibits a parabolic shape. The displacement–time curves of the ball along the x-axis and y-axis are plotted in Fig. 2.9b, c, respectively. Figure 2.9b describes the ball displacement increases linearly with time along the x-axis. Figure 2.9c indicates the displacement–time curve exhibits a parabolic shape along the y-axis and it increases with time, and then decreases with time.

Theoretical Calculation: Hroncová and Grieš establish a mathematical model of a particle moving in x–y space within the homogeneous air resistance environment integrating classic mechanics and aerodynamics. The motions of the particle projectile in the horizontal x-direction and vertical y-direction are given as Eqs. 2.3.2a, 2.3.2b, respectively.

$$x = \frac{v_0}{k_1}\cos\theta_0\left(1 - e^{-k_1 t}\right) \tag{2.3.2a}$$

$$y = \frac{g + v_0 k_1 \sin\theta_0}{k_1}\left(1 - e^{-k_1 t}\right) - \frac{g}{k_1}t \tag{2.3.2b}$$

where v_0 is the particle initial velocity, θ_0 is the particle initial projectile angle, $k_1 = F/mv$, F is the air resistance also called aerodynamic drag F_d (see Eq. 2.4 for calculation), v is the velocity of the particle, and g is the gravitational acceleration.

In their article, a case study is conducted using Eqs. 2.3.2a and 2.3.2b to find a particle displacement versus time in a Cartesian coordinate system o(x, y). The general information: physical property and initial parameters, is listed in Table 2.2. They are: coefficient K_d = 0.00002 kg/m, initial velocity v_0 = 46.5 m/s, initial projectile angle θ_0 = 45°.

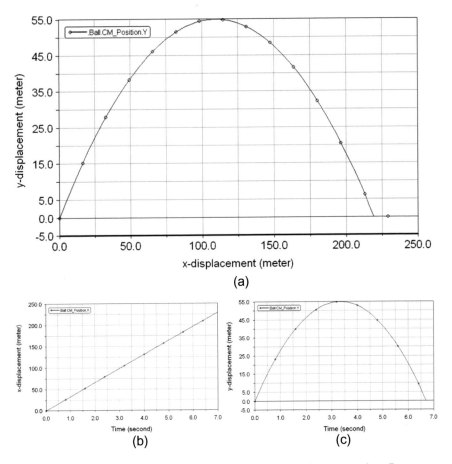

Fig. 2.9 Dynamic simulation of the soccer ball projectile motion versus time 7 s. **a** x–y displacement, **b** x–t displacement, **c** y–t displacement

Table 2.2 Physical property and initial parameters

Physical property	Initial parameters	
Coefficient	Initial velocity	Initial projectile angle
K_d (kg/m)	v_0 (m/s)	θ_0 (°)
0.00002	46.5	45

For the comparison of two models, here, the displacements of the particle versus time 7 s are calculated and plotted in Fig. 2.10a–c. The particle is shot from an initial point o(0, 0) and to the ground at point (219, 0). Figure 2.10a plots its x–y displacement in coordinate system o(x, y). The particle flight trajectory has a parabolic shape. The displacement–time curves in the x-direction and y-direction are given in Fig. 2.10b and c, respectively. Figure 2.10b shows the displacement increases

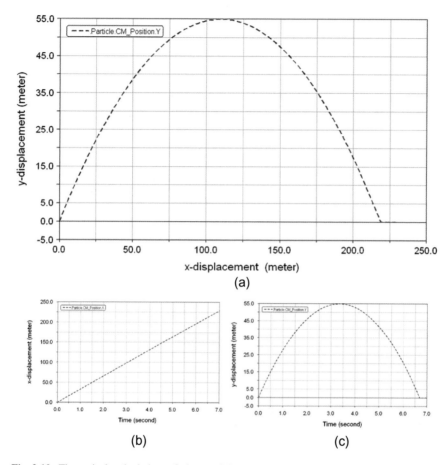

Fig. 2.10 Theoretical calculation of the particle projectile motion versus time 7 s. **a** x–y displacement, **b** x–t displacement, **c** y–t displacement

linearly with time in the x-direction. Figure 2.10c indicates the displacement–time curve exhibits a parabolic shape in the y-direction and it increases with time, and then decreases with time.

Comparison of the projectile motions of the soccer ball and particle versus time 7 s is given in Fig. 2.11a–c. It notes that even though two displacement results are conducted using two different models from two different teams, the x–y, x-t and y–t displacements compare favorably with each other. Therefore, there is no difference between the simulated x–y displacements and calculated x–y displacements. No matter what kind of result it is, the flight path is un-symmetric about its mid point due to aerodynamic drag. The displacement in the x–y plane shows a parabolic shape, in which y increases from 0 to 55 m within x from 0 to 108 m and then y decreases from 55 to 0 m within x from 108 to 219 as shown in Fig. 2.11a. The x-displacement linearly increases from 0 to 219 m within the time range of 0–6.7 s as shown in Fig. 2.11b. The y-displacement shows a parabolic shape, which increases from 0 to

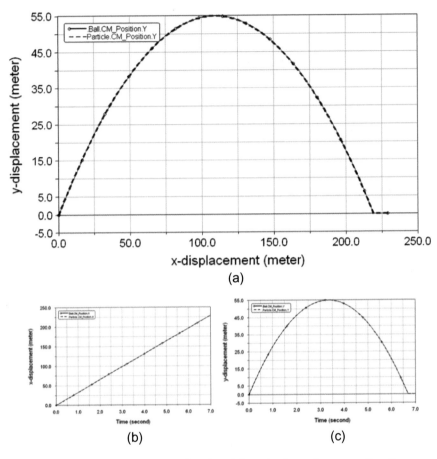

Fig. 2.11 Comparison of the soccer ball and particle projectile motion versus time 7 s. **a** x–y displacemen, **b** x–t displacement (Li and Li 2019), **c** y–t displacement (Li and Li 2019)

55 m within 0 to 3.3 s and then decreases from 55 to 0 m within 3.3 to 6.7 s as shown in Fig. 2.11c. It is important to note that the max projectile length is 226 m and the max projectile height is 55 m.

This analysis case validates the potential of the dynamics model to predict a soccer ball projectile motion with aerodynamic drag (air resistance). The analysis results identify that the dynamics simulation will give a higher confidence in the soccer ball kinematics simulation with the consideration of the air resistance. The analysis results also illustrate that the dynamics model can capture the ball projectile motion such as the horizontally shooting range (x-displacement) and vertically shooting high (y-displacement). Anyway, for the 3D dynamics model created in this research, it can also capture the motion in the z-direction.

In reality, a soccer ball kicked by a player will not be expected to shoot at such a great speed of 46.5 m/s and move so far away from 219 m. The moving direction of the ball can be random in 3D space instead of 2D space because a player can choose to kick at any position and orientation around the ball. For addressing these issues, some case studies are demonstrated in the next four chapters.

References

Asai T, Seo K, Kobayashi O, Sakashita R (2007) Fundamental aerodynamics of the soccer ball. Sports Eng 10:101–109

Asai T, Seo K (2013) Aerodynamic drag of modern soccer balls. Springerplus 2:171

Atkins PW, De Paula J (2008) Kurzlehrbuch Physikalische Chemie, 4th edn. Wiley-VCH, Weinheim

Bray K, Kerwin DG (2003) Modelling the flight of a soccer ball in a direct free kick. J Sports Sci 21:75–85

Christenson A, Cao P, Tang J (2018) Optimal football pressure as a function of a footballer's physical abilities. Presented at the 12th conference of the international sports engineering association, Brisbane, Queensland, Australia, 26–29 March

Goff JE, Carré MJ (2009) Trajectory analysis of a soccer ball. Am J Phys 77:1020–1027

Hroncová D, Grieš M (2014) Trajectories of projectiles launched at different elevation angles and modify design variable in MSC Adams/View. Appl Mech Mater 611:198–207, ISSN: 1662–7482

Li Y, Li Q (2019) Soccer ball spatial kinematics and dynamics simulation for efficient sports analysis. Asian J Adv Res Reports 7(4):1–18, ISSN: 2582–3248

Li Y, Liu WY (2014) Comparison of advanced dragline dynamics models for efficient engineering analysis. Int J Adv Manuf Technol 72(5–8):757–764

Maczynski A, Szczotka M (2002) Comparison of models for dynamic analysis of a mobile telescope crane. J Theor Appl Mech 40(4):1051–1074

Meriam JL, Kraige LG (2002) Engineering mechanics, dynamics, 5th edition. Wiley, Incorporated, New York, USA

Smith MR, Hilton DK, Van Sciver SW (1999) Observed drag crisis on a sphere in flowing He I and He II. Phys Fluids 11:751–753

Synge JL, Griffith BA (1959) Principles of mechanics. McGraw Hill, New York

Chapter 3
Spatial Kinematics and Dynamics Simulation of Soccer Ball Projectile Motion

3.1 Overview

In this Chapter, a spatial two-body dynamics model of soccer field-ball has been developed. The model involves gravity, aerodynamic drag, and contact force. The trajectory of the ball flight and the force applied to the ball are visualized through virtual kinematics and dynamics simulation. The time-varying displacement and time-varying force are calculated. The spatial kinematics and dynamics performance of the ball are analyzed. The results show the max projectile height and range, or kicking force increase with the increase of the initial velocity. All works have built foundations to predict the soccer ball targets through dynamics modeling, optimization design, virtual simulation.

3.2 Initial Configuration

In our early investigation (Li and Li 2019), a simple dynamics modeling has been developed for the dynamic simulation of a soccer ball projectile motion. In this Chapter, the study has been extended to improve the model and presented more detail explanation. Figure 3.1 shows the dynamics modeling and virtual simulation of a soccer ball projectile motion. The model is created by ignoring the ball flying spin, Magnus effect, gyroscopic moment, and acceleration of Coriolis. A Cartesian coordinate system o(x, y, z) is attached to the soccer field at the center of the soccer ball, where x–z plane is attached on the field with x-axis along the length direction and z-axis along the width direction, and y-axis is perpendicular to the x–z plane.

Table 3.1 summarizes the geometric parameters of field-ball and the physical properties of environment. They are a virtual soccer field in length LF = 120 m and width WF = 90 m. The mass of the ball m = 0.43 kg, its diameter d = 0.2286 m, sea level static density ρ = 1.221 kg/m^3, and drag coefficient C_d = 0.25 (NASA 2018).

© The Author(s), under exclusive license to Springer Nature Singapore Pte Ltd. 2022
Y. Li, *Motion Analysis of Soccer Ball*,
SpringerBriefs in Applied Sciences and Technology,
https://doi.org/10.1007/978-981-16-8652-8_3

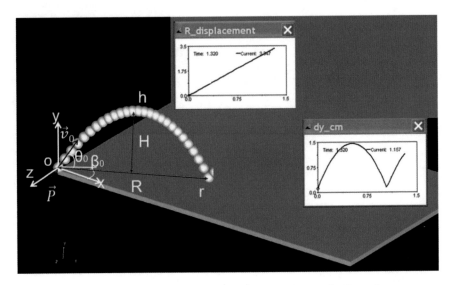

Fig. 3.1   Dynamics modeling and virtual simulation of a soccer ball projectile motion

Table 3.1   Geometric parameters and physical properties

Geometric parameters				Physical properties	
Field		Ball		Environment	
Length	Width	Diameter	Mass	Air density	Drag coefficient
LF (m)	WF (m)	d (m)	m (kg)	ρ (kg/m$^{3)}$)	C_d (kg/m)
120	90	0.2286	0.43	1.221	0.25

Figure 3.2 shows a free-body diagram of the initial configuration of kicking force and initial velocity vector. A soccer ball is initially at rest on the field and a player kicks the ball at its center with force $\overrightarrow{P}$. When the ball leaves the player's foot, the ball travels with initial velocity $\overrightarrow{v}_0$. If the foot and ball are in contact with time Δt, the relationship between the force vector $\overrightarrow{P}$ and velocity vector $\overrightarrow{v}_0$ can be established as Eq. 3.1 by employing the principle of linear impulse and momentum (Meriam and Kraige 2002). Equation 3.1 can be written as Eqs. 3.1a, 3.1b.

$$\overrightarrow{P}\,\Delta t = m\,\overrightarrow{v}_0 \tag{3.1}$$

$$\overrightarrow{P} = \frac{m\,\overrightarrow{v}_0}{\Delta t} \tag{3.1a}$$

$$\overrightarrow{v}_0 = \frac{\overrightarrow{P}\,\Delta t}{m} \tag{3.1b}$$

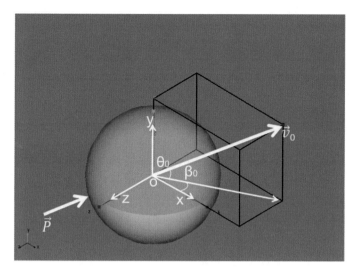

Fig. 3.2 Free-body diagram of a soccer ball with a kicking force and initial velocity

In this Chapter, two problems need to be solved (i) using Eq. 3.1a to find kicking force $\overrightarrow{P}$ for a given $\overrightarrow{v}_0$; and (ii) reversely, using Eq. 3.1b to find $\overrightarrow{v}_0$ for a given kicking force $\overrightarrow{P}$.

The ball is launched with initial velocity v_0 at initial projectile angle $\theta_0 = 30°$, and initial orientation angle $\beta_0 = 30°$. Initial velocity vector $\overrightarrow{v}_0$ can be separated into three scalar components v_{x0}, v_{y0}, and v_{z0} in x–y–z directions as expressed in Eqs. 3.2a, 3.2c.

$$v_{x0} = v_0 \cos\theta_0 \cos\beta_0 \tag{3.2a}$$

$$v_{y0} = v_0 \sin\theta_0 \tag{3.2b}$$

$$v_{z0} = -v_0 \cos\theta_0 \sin\beta_0 \tag{3.2c}$$

Submitting initial projectile angle $\theta_0 = 30°$ and initial orientation angle $\beta_0 = 30°$ into Eqs. 3.2a–3.2c, the initial velocity components v_{x0}, v_{y0}, and v_{z0} can given as Eqs. 3.3a, 3.3c.

$$v_{x0} = 0.75 v_0 \tag{3.3a}$$

$$v_{y0} = 0.5 v_0 \tag{3.3b}$$

$$v_{z0} = -0.433 v_0 \tag{3.3c}$$

3.3 Dynamics Modeling

Figure 3.3 shows two free-body diagrams of the ball in two instantaneous positions, where the ball moves up and down. The model is created by assuming the air around the ball is homogenous, and ignoring the ball flying spin, Magnus effect, drag moment, gyroscopic moment, and acceleration of Coriolis. The forces applied on the ball include gravity $\vec{G}$ and aerodynamic drag $\vec{F}_d$. When the ball touches on field, contact force $\vec{F}_n$ occurs between the two bodies. The equations of three forces are built in below. Some equations have been derived in Chap. 2 and are repeated in this Chapter for the sake of completeness.

Gravity: gravity $\vec{G}$ acts to the ball in negative y-direction as shown in Fig. 3.3. If g = 9.8 m/s^2 denotes gravitational acceleration, it can be calculated by Eq. 3.4.

$$\vec{G} = -mg \tag{3.4}$$

Aerodynamic Drag: during traveling, the air resists the ball moving and applies aerodynamic drag $\vec{F}_d$ on the ball. As indicated in Chap. 2, the aerodynamic drag can be calculated by Eqs. 3.5 and 3.6 (Bray and Kerwin 2003; Asai et al. 2007, 2013; Goff and Carré 2009).

$$\vec{F}_d = -\frac{1}{2}\rho C_d A \vec{v} \lfloor \vec{v} \rfloor = -K_d \vec{v} |\vec{v}| \tag{3.5}$$

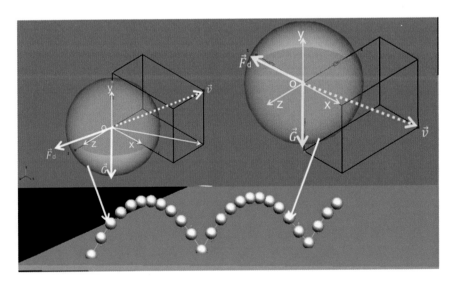

Fig. 3.3 Soccer ball free-body diagrams at two instantaneous positions

$$K_d = \frac{1}{8}\rho C_d \pi d^2 \qquad (3.6)$$

where, ρ is the density of the air, C_d is the drag coefficient, A *is* the cross-sectional area of the ball, $\overrightarrow{v}$ is the velocity of the ball flight, and d is the diameter of the ball. The drag coefficients of a soccer ball are suggested between 0.1 and 0.5 (Smith et al. 1999). It depends on the boundary conditions, such as the surface roughness of a ball, as well as the laminar flow or turbulence of the atmospheric layer.

Since assuming that the air around the ball is homogenous, it will have coefficient $K_{dx} = K_{dy} = K_{dz} = K_d = 0.00622$ kg/m. Then, the aerodynamic drag $\overrightarrow{F}_d$, which has built in Chap. 2, is simplified to Eqs. 3.7a, 3.7c. Therefore, vector Eq. 3.5 can be written as three sets of scalar Equations to calculate the component forces F_{dx}, F_{dy} and F_{dz}, in x–y–z directions, respectively.

$$F_{dx} = -K_d \dot{x}|\dot{x}| \qquad (3.7a)$$

$$F_{dy} = -K_d \dot{y}|\dot{y}| \qquad (3.7b)$$

$$F_{dz} = -K_d \dot{z}|\dot{z}| \qquad (3.7c)$$

Contact Force: Figure 3.4 shows the configurations of the ball-field contact, and the contact model is built as the mass-dumping-stiffness system. Where c donates the contact damping coefficient and $\overrightarrow{v}$ is the velocity vector, and then the damping force vector $\overrightarrow{F}_c$ is given as

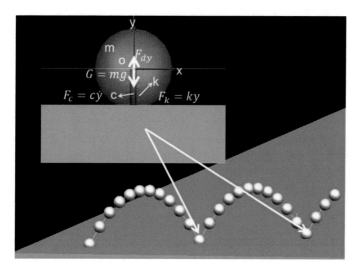

Fig. 3.4 A ball-field contact model with a mass-dumping-stiffness system

$$\overrightarrow{F}_c = c\overrightarrow{v} \tag{3.8a}$$

Also, k donates the contact stiffness and $\overrightarrow{\delta}$ is the displacement vector, and then the elastic force vector $\overrightarrow{F}_k$ is given as

$$\overrightarrow{F}_k = k\overrightarrow{\delta} \tag{3.8b}$$

where, the contact damping coefficient $c = 3.8 \times 10^5$ N/m and contact stiffness $k = 97.2$ N.s/m. They cannot find in material property handbooks. The data given here are calculated results. The contact force is established as Eq. 3.9 by employing Newtown's Second Law and Hooke's law.

$$\overrightarrow{F}_n = \overrightarrow{F}_c + \overrightarrow{F}_k \tag{3.9}$$

Substituting from Eqs. 3.8a, 3.8b to 3.9, the contact force is represented as Eq. 3.10a

$$\overrightarrow{F}_n = c\overrightarrow{v} + k\overrightarrow{\delta} \tag{3.10a}$$

Since contact force $\overrightarrow{F}_n$ only acts in y-direction, it can be expressed as F_{ny} and given in Eq. 3.10b.

$$F_{ny} = c\dot{y} + ky \tag{3.10b}$$

General Dynamics Equation: submitting gravity $\overrightarrow{G}$, aerodynamic drag $\overrightarrow{F}_d$ and contact force $\overrightarrow{F}_{ny}$ into Newton Equation, general dynamics model of soccer ball can be written as

$$\sum \overrightarrow{N} = m\overrightarrow{a} = \overrightarrow{G} + \overrightarrow{F}_d + e\overrightarrow{F}_{ny} \tag{3.11}$$

where, $e = 0$, when $\overrightarrow{\delta} >= d/2$, there is no penetration between the ball and field, and the contact force is zero; and $e = 1$, when $\overrightarrow{\delta} < d/2$, there is penetration between the ball and field, and the contact force is larger than zero.

Separating Eq. 3.11 into three components in x–y–z directions, the dynamics model is expressed as differential Eqs. 3.12a, 3.12c.

$$m\ddot{x} = -K_d\dot{x}|\dot{x}| \tag{3.12a}$$

$$m\ddot{y} = -mg - K_d\dot{y}|\dot{y}| + e(c\dot{y} + ky) \tag{3.12b}$$

$$m\ddot{z} = -K_d\dot{z}|\dot{z}| \tag{3.12c}$$

$$
\begin{bmatrix}
1 & & & & & \\
 & 1 & & & & \\
 & & 1 & & & \\
 & & & m & & \\
 & & & & m & \\
 & & & & & m
\end{bmatrix}
\begin{bmatrix}
F_{dx} \\
F_{dy} \\
F_{dz} \\
\ddot{x} \\
\ddot{y} \\
\ddot{z}
\end{bmatrix}
=
\begin{bmatrix}
-K_d\dot{x}|\dot{x}| \\
-K_d\dot{y}|\dot{y}| \\
-K_d\dot{z}|\dot{z}| \\
-K_d\dot{x}|\dot{x}| \\
-mg - K_d\dot{y}|\dot{y}| + e(c\dot{y} + ky) \\
-K_d\dot{z}|\dot{z}|
\end{bmatrix}
\tag{3.13}
$$

The dynamics model is created using the simultaneous constraint method (Gardner 2001) for the simulation of constrained soccer ball motion system. Assembling Eqs. 3.7a, 3.7c and 3.12a, 3.12c in a matrix from, the general dynamics model is built as Eq. 3.13. This equation forms a system of six equations and six unknowns, F_{dx}, F_{dy}, F_{dz}, $\ddot{x}$, $\ddot{y}$, and $\ddot{z}$. When embedding this matrix equation in a simulation system of soccer ball motion, if the accelerations, are integrated, then velocities$\dot{x}$, $\dot{y}$ and $\dot{z}$ and displacements, x, y, and z, will be available to compute the matrix and the right side of the equation. Figure 3.5 shows the dynamic simulation diagram of this system. There is no input system. The initial conditions will determine its response.

So far, the initial configuration and dynamics modeling are all obtained. Then, the work turns to simulate and analyze the soccer ball projectile motion.

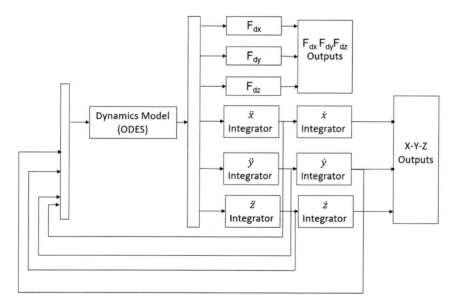

Fig. 3.5 Dynamic simulation diagram of the soccer ball projectile motion

3.4   Kinematics and Dynamics Simulation and Results Analysis

The case studies have been derived in our previous studies (Li and Li 2019). In this Chapter, the case is employed for the further explanation of the application. The kinematics and dynamics simulations of a soccer ball projectile motion are performed by selecting four groups of initial velocities and simulating times. The results are output as animation graphics and numerical diagrams. The flight trajectories of the soccer ball are visualized in the virtual environment of a soccer field. The curves of displacement–time and force–time are measured by assigning four initial velocities $v_0 = 5, 10, 15$, and 20 m/s, respectively. The spatial kinematics and dynamics results are output as a report and analyzed.

3.4.1   Kinematics Simulation and Results Analysis

Figure 3.6a–d shows dynamic simulation of a soccer ball projectile motion with four initial velocities $v_0 = 5, 10, 15$, and 20 m/s responding to four simulating times t = 0.65, 1.2, 1.7, and 2.5 s, respectively (Li and Li 2019). The animations exhibit that four graphics of soccer trajectories all show parabolic shapes. The curves attached

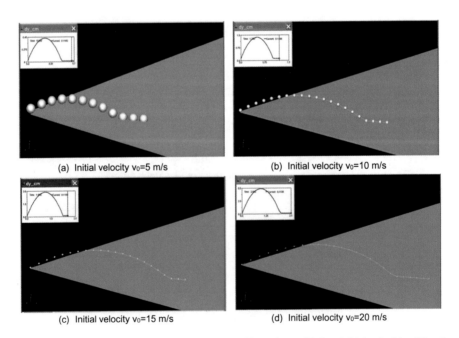

(a) Initial velocity v_0=5 m/s

(b) Initial velocity v_0=10 m/s

(c) Initial velocity v_0=15 m/s

(d) Initial velocity v_0=20 m/s

Fig. 3.6 Dynamic simulations of soccer ball projectile motions with four initial velocities (Li and Li 2019)

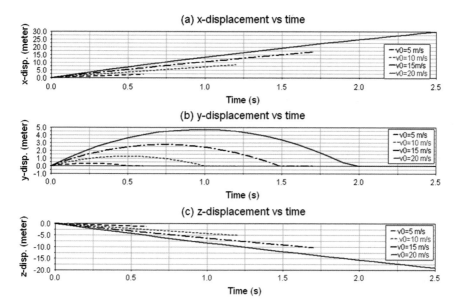

Fig. 3.7 The displacement–time curves of soccer ball in x–y–z directions with four initial velocities $v_0 = 5, 10, 15,$ and 20 m/s (Li and Li 2019)

on the Figures display the flight heights vs the time. It can be found the bigger the initial velocity, the higher the ball reaches, and the farther the ball flies.

Figure 3.7a–c plots the displacement–time curves of soccer ball in x–y–z directions with the four groups of initial velocities v_0=5, 10, 15, and 20 m/s responding to the four sets of times t = 0.65, 1.2, 1.7, and 2.5 s, respectively. They reveal the instantaneous positions of the ball traveling in x–y–z directions. Based on the data in Fig. 3.7a–c, the positions of the ball are expressed in 3D space in coordinate o(x, y, z) as shown in Fig. 3.8a–d.

Figures 3.7 and 3.8 provide some information about the ball flight parameters. The max projectile height H and the max projectile range R are two key components concerned in this Chapter. The max projectile height H is defined as the distance between the initial point (x_0, y_0, z_0) and the highest point h (x_H, y_H, z_H) as shown in Fig. 3.1. For example, in Fig. 3.8d it can find that the max projectile height of H = 4.75 m occurs at point h, $y_H = 4.75$, $x_H = 13.58$ m, and $z_H = -8.16$ m, when $v_0 = 20$ m/s. Figure 3.7b illustrates the max projectile height of H = 4.75 m happens at time t = 1 s.

Figure 3.1 also indicates the parameter of the max projectile range R, which is defined as the distance between the initial point (x_0, y_0, z_0) and landing point r (x_R, y_R, z_R). Therefore, the mathematical expression of the max projectile range R can be written in Eq. 3.14.

$$R = \sqrt{(x_R - x_0)^2 + (z_R - z_0)^2} \qquad (3.14)$$

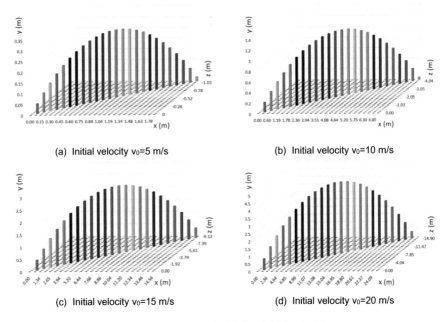

Fig. 3.8 Instantaneous positions of soccer ball with four initial velocities

For example, in Fig. 3.8d, when the ball lands in the field on x–z plane, y_H equals to 0 m. The max x-projectile range equals the distance between $x_0 = 0$ and $x_R = 24.4.23$ m and the max z-projectile range equals the distance between $z_0 = 0$ and $z_R = -15.46$ m. Using Eq. 3.14, it can get the max projectile range R = 29.33 m. In Fig. 3.7b, it notes for $y_H = 0$, time t = 2 s.

From this example, it can find the flight height and distance of a soccer ball when kicking a soccer ball with an initial velocity $v_0 = 20$ m/s at initial projectile angle $\theta_0 = 30°$, and initial orientation angle $\beta_0 = 30°$. They are the max projectile height H = 4.75 m occurs at the point h (13.58, 4.75, −8.16) at time t = 1 s and the max projectile range R = 29.33 m at the point r (24.93, 0, -15.46) at time t = 2 s.

Similarly, the max projectile height H and projectile range R for the initial velocities $v_0 = 5$, 10, and 15 m/s can be obtained. Table 3.2 lists the max projectile height

Table 3.2 Max projectile height H for initial velocities v_0=5, 10, 15, and 20 m/s	Initial velocity	Max projectile height		
	v_0 (m/s)	H (m)	@ Time t (s)	@ Point h (x_H, y_H, z_H) (m)
	5	0.32	0.26	(0.97,0.32,-0.56)
	10	1.25	0.50	(3.65,1.25,-2.13)
	15	2.76	0.74	(7.86,2.76,-4.65)
	20	4.75	1.00	(13.58,4.75,-8.16)

Table 3.3 Max projectile range R for initial velocities v_0=5, 10, 15, and 20 m/s

Initial velocity	Max projectile range		
v_0 (m/s)	R (m)	@ Time (s)	@ Point r (x_R, y_R, z_R) (m)
5	2.14	0.5	(1.85,0,-1.07)
10	8.27	1.0	(7.12,0,-4.20)
15	17.64	1.5	(15.1,0,-9.12)
20	29.33	2.0	(24.93,0,-15.46)

H for initial velocities v_0=5, 10, 15, and 20 m/s. Table 3.3 lists the max projectile range R for initial velocities v_0=5, 10, 15, and 20 m/s. In Tables 3.2 and 3.3, some significant results about the soccer ball kinematics can be obtained and presented in the following points.

1. It can figure out the variations of the max projectile height H or the max projectile range R with an initial velocity v_0. Table 3.2 presents four sets of initial velocities and the max projectile height H. The relationship between the max projectile height H and initial velocity v_0 is described in Fig. 3.9. The max projectile height H non-linearly increases with the increase of v_0. Table 3.3 presents four sets of initial velocities and the max projectile range R. The relationship between the max projectile range R and initial velocity v_0 is depicted in Fig. 3.10. Similarly, it can find that the max projectile range R increases with the increase of v_0 non-linearly. Therefore, it can conclude that if launching a ball with a series of initial velocities at the same initial projectile angle and initial orientation angle, the max projectile height H or the max projectile range R non-linearly increases with the increase of initial velocity.

2. It can figure out the variations of the landing point of a soccer ball with an initial velocity v_0. Table 3.3 presents four sets of initial velocities corresponding to the position of the four landing points. Figure 3.11 plots the distribution of landing

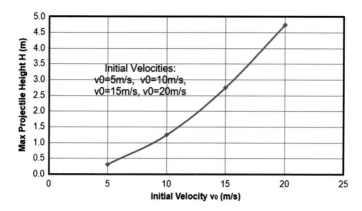

Fig. 3.9 The variation of the max projectile height H vs initial velocity v_0

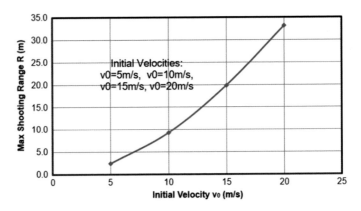

Fig. 3.10 The variation of the max projectile range R vs initial velocity v_0

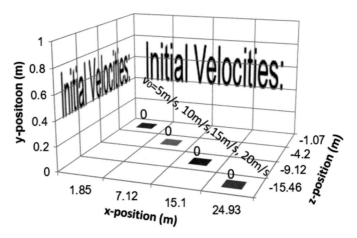

Fig. 3.11 The distribution of the landing points in x–y–z space for initial velocities $v_0 = 5, 10, 15,$ and 20 m/s

points in the x–y–z space with the initial velocities v_0=5, 10, 15, and 20 m/s. The four landing points are positioned in field located on the x–z plane (where y = 0). Furthermore, in Table 3.3, the coordinate data of the four landing points x_R and z_R reveal the distribution of the four landing points in field on x–z plane as plotted in Fig. 3.12. The result indicates that four landing points are distributed in a straight line under the condition of the initial velocities v_0=5, 10, 15, and 20 m/s, the initial projectile angle $\theta_0 = 30°$, and initial orientation angle $\beta_0 = 30°$. Therefore, it can conclude that if launching a ball with a series of initial velocities at the same initial projectile angle and initial orientation angle, the landing points present a straight-line distribution on the field.

3. It can figure out the variations of the max projectile height H with its max projectile range R_H for the four sets of initial velocities v_0=5, 10, 15, and 20 m/s.

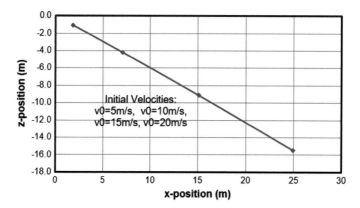

Fig. 3.12 The distribution of the landing points in x–z plane for initial velocities $v_0 = 5, 10, 15,$ and 20 m/s

Table 3.2 presents four sets of initial velocities, the max projectile height H and their points. Figure 3.13 shows the distribution of four points of the max projectile height in the coordinate x–y-z. And, the corresponding projectile range R_H is projected on the x–z plane. R_H is defined as the distance measured from the initial point (x_0, z_0) to point (x_H, z_H) and calculated by Eq. 3.15.

$$R_H = \sqrt{(x_H - x_0)^2 + (z_H - z_0)^2} \tag{3.15}$$

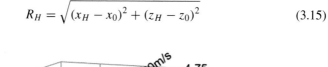

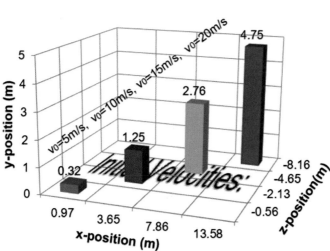

Fig. 3.13 The distribution of the max projectile height points in x–y-z space for initial velocities $v_0 = 5, 10, 15,$ and 20 m/s

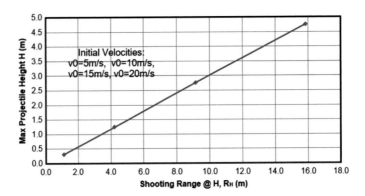

Fig. 3.14 The relationship between the max projectile height H and its projectile range R_H for initial velocities $v_0 = 5, 10, 15$, and 20 m/s

Substituting x_H and z_H data listed in Table 3.2 to Eq. 3.15, and then it can get R_H. The relationship between the max projectile height H and the range R_H is plotted in Fig. 3.14. The max projectile height H rises with the rising of the range R_H linearly under the condition of initial velocities v_0=5, 10, 15, and 20 m/s, initial projectile angle $\theta_0 = 30°$, and initial orientation angle $\beta_0 = 30°$. Thus, it can conclude that if launching a ball with a series of initial velocities at the same initial projectile angle and initial orientation angle, the max height points that the ball arrives form a straight-line distribution in space.

3.4.2   Dynamics Simulation and Results Analysis

The kinematics simulation lays a foundation for the dynamics simulation of the soccer ball projectile motion. The dynamics studies focus on finding the kicking force, animating the flight trajectory, and outputting the dynamic force. The detailed studies are focused on the following two aspects.

1.   The Impulsive Force of Kicking a Soccer Ball

The case studies of the kinematics simulation have selected four initial velocities v_0 = 5, 10, 15, and 20 m/s. In this Section, the problem is to solve how big kicking force $\overrightarrow{P}$ is, for a given $\overrightarrow{v}_0$. Figure 3.15 indicates a diagram of initial configuration of soccer ball. The problem states that ball is initially at rest on the field and a player kicks it at the center of the ball. When the ball leaves the player's foot, the ball flies with initial velocity v_0 at initial projectile angle $\theta_0 = 30°$, and initial orientation angle $\beta_0 = 30°$. If the contacting time between foot and ball is $\Delta t = 0.012$ s, the average force P can be calculated using the principle of linear impulse and momentum as presented in Eq. 3.1a.

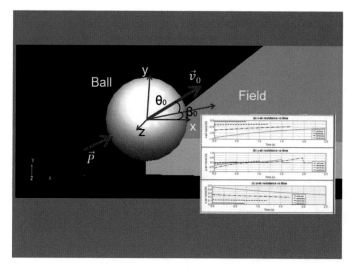

Fig. 3.15 Initial configuration of kicking a soccer ball

The magnitude and direction of the force can be obtained, individually. In Eq. 3.1a, it can be seen that the linear impulsive force applied on the ball is equal to the change of the linear momentum. The equation builds a linear relationship between the impulsive force P and the initial velocity v_0. Figure 3.16 demonstrates the variation of P with v_0. The results show that the impulsive force P linearly increases with the increase of the initial velocity v_0. When applying initial velocities $v_0 = 5, 10, 15$, and 20 m/s on the soccer ball at initial projectile angle $\theta_0 = 30°$ and initial orientation angle $\beta_0 = 30°$, the average impulsive forces are P = 107.5, 215.0, 322.5 and 430 N, individually.

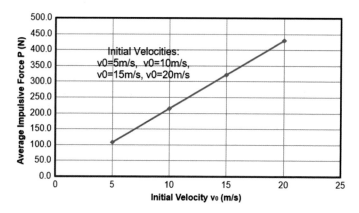

Fig. 3.16 The variation of average impulsive force P with four initial velocities $v_0 = 5, 10, 15$, and 20 m/s

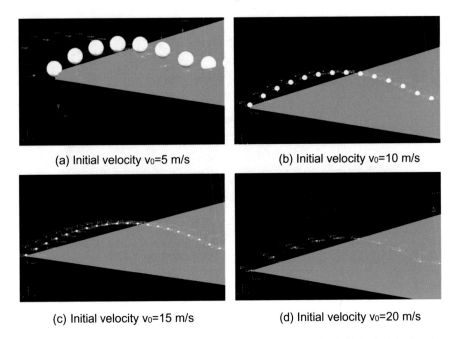

(a) Initial velocity v₀=5 m/s   (b) Initial velocity v₀=10 m/s

(c) Initial velocity v₀=15 m/s   (d) Initial velocity v₀=20 m/s

Fig. 3.17 Dynamic simulations of soccer ball projectile motions with four initial velocities showing successive aerodynamic drag graphics (Li and Li 2019)

2.   The Aerodynamic Drag of Applying on a Soccer Ball

Figure 3.17a–d shows the four sets of projectile motions of soccer ball with four initial velocities $v_0 = 5, 10, 15$, and 20 m/s respect to the times t = 0.7, 1.2, 1.7, and 2.5 s, respectively. Successive aerodynamic drag graphics are shown up by using the virtual simulation method. The instantaneous aerodynamic drag is separated into x–y-z components in the series of the dynamic local coordinate systems. The individual component is changed with the instantaneous position and time.

Figure 3.18a–c plots the four sets of the aerodynamic-drags applied on the ball in x–y-z directions with respect to four initial velocities $v_0 = 5, 10, 15$, and 20 m/s, corresponding with times 0.65, 1.2, 1.7, and 2.5 s, respectively. The four charts reveal the distributions of aerodynamic drag vs the initial velocity or aerodynamic drag vs time. Some significant results about the soccer ball dynamics can be obtained and presented in the following points.

Figure 3.18a shows the four sets of the curves of the aerodynamic drag-time history along the x-direction. For a given time, the aerodynamic drag in x-direction $|F_{dx}|$ increases with the increase of the initial velocity; and for a given trajectory, the aerodynamic drag in x-direction $|F_{dx}|$ decreases with the increase of the time. This is because when the soccer ball traveling through the air, the air around the ball generates resistance called aerodynamic drag or air resistance. The aerodynamic drag in x-direction resists the ball moving forward and causes the velocity of the ball flight slowing down in x-direction. In this case, Eq. 3.7a can be expressed as $|F_{dx}| = K_d v_x^2$.

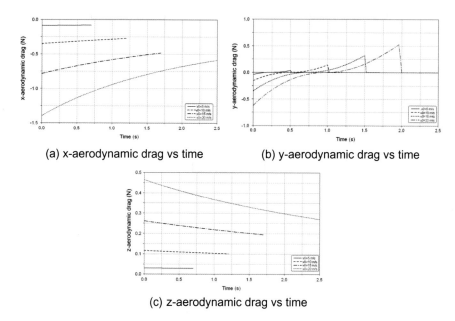

(a) x-aerodynamic drag vs time   (b) y-aerodynamic drag vs time

(c) z-aerodynamic drag vs time

Fig. 3.18 Four sets of aerodynamic-drags applied on the soccer ball in x–y–z directions with four initial velocities $v_0 = 5$, 10, 15, and 20 m/s

It notes that the aerodynamic drag $|F_{dx}|$ is proportional to the squared velocity of the ball v_x^2. Therefore, the aerodynamic drag $|F_{dx}|$ decreases with the decrease of the velocity.

Figure 3.18b shows the four sets of the curves of the aerodynamic drag-time history along the y-direction. For a given trajectory, the aerodynamic drag F_{dy} rises from the lowest point of the negative value through zero to the highest point of the positive value. This is because when the ball flying through the air in the y-direction, both gravity and aerodynamic drag apply on the ball in negative or positive y-direction. During the ball rising stage, the gravity and aerodynamic drag resist the ball moving up makes it slow down. Equation 3.7b can be written as $|F_{dy}| = K_d v_y^2$. It notes that the aerodynamic drag $|F_{dy}|$ is proportional to the squared velocity of the ball v_y^2. Therefore, the aerodynamic drag $|F_{dy}|$ decreases with the decrease of the velocity until the ball reaches the highest point h, where the velocity of the ball equals zero. Then, the ball reverses its y-direction and goes down to the field. During the moving down stage, the velocity of the ball increase with the increase of the time. The gravity drags the ball moving down but the aerodynamic drag keeps on resisting the ball moving down. The aerodynamic drag $|F_{dy}|$ increases with the increase of the velocity. At the end of the flight, the aerodynamic drag drops to zero because the ball lands on the field.

Figure 3.18c shows the four sets of the curves of aerodynamic drag-time history along the z-direction. These curves have similar trends as the curves along the x-direction. For a given time, the aerodynamic drag $|F_{dz}|$ increases with the increase of the initial velocity; and for a given trajectory, the aerodynamic drag decreases with the increase of the time. The interpretation of this variation is also similar to the interpretation in Fig. 3.18a. This is because when the ball moving through the air in z-direction, the air resists the ball moving in. The aerodynamic drag the velocity in z-direction causes the ball slowing down. Here, Eq. 3.7c can be written as $|F_{dz}| = K_d v_z^2$. It notes that the aerodynamic drag $|F_{dz}|$ is proportional to the squared z-velocity of the ball v_z^2. Therefore, the aerodynamic drag $|F_{dz}|$ decreases with the decrease of the velocity.

This study employs a simple case to indicate the application of the proposed dynamics model on the kinematics and dynamics simulation of a soccer ball projectile motion. The results of the soccer ball motion and aerodynamic drag are presented as 3D animation graphics and numerical diagrams. The analysis is focused on the influence of initial velocity on the max projectile height and range. The effect of an aerodynamic drag on a ball flight trajectory also is addressed. The impulsive force of kicking a soccer ball is calculated and the initial velocity produced by the impulsive force is defined by giving an initial projectile angle and an initial orientation angle.

References

Asai T, Seo K (2013) Aerodynamic drag of modern soccer balls. Springerplus 2:171

Asai T, Seo K, Kobayashi O, Sakashita R (2007) Fundamental aerodynamics of the soccer ball. Sports Eng 10:101–109

Bray K, Kerwin DG (2003) Modelling the flight of a soccer ball in a direct free kick. J Sports Sci 21:75–85

Goff JE, Carré MJ (2009) Trajectory analysis of a soccer ball. Am J Phys 77:1020–1027

Gardner J (2001) Simulations of machines using Matlab and Simulink, 1st edn. University of California, Brooks

Li Y, Li Q (2019) Soccer ball spatial kinematics and dynamics simulation for efficient sports analysis. Asian J Adv Res Rep 7(4):1–18, ISSN: 2582-3248

Meriam JL, Kraige LG (2002) Engineering mechanics, dynamics, 5th edition. Wiley, Incorporated, New York, USA

NASA (2018) Drag on a Soccer Ball. National Aeronautics and Space Administration Technic Paper, www.nasa.gov

Smith MR, Hilton DK, Van Sciver SW (1999) Observed drag crisis on a sphere in flowing He I and He II. Phys Fluids 11:751–753

Chapter 4
Optimization Design of Soccer Ball Flight Trajectory

4.1 Overview

The intelligent analysis of the motion of a soccer ball requires accurately predicting its trajectory and finding the accurate design parameters to achieve a goal. The optimization design can efficiently improve the design parameters within the required range to exact evaluation of flight trajectory. In this Chapter, an optimization method is proposed to plan, evaluate, and optimize the traveling trajectory of a soccer ball. The parameterization technology has been integrated into the optimization design for automatically capturing the expected target, which is expressed as a function of all design parameters. A case study goes through multi-body dynamics modeling, dynamic simulation, and optimal objective modeling. A soccer ball shooting at target has been predicted. The result of optimization design has given the most optimal combination of the design parameters including the initial velocity, initial projectile angle, and initial orientation angle. This research provides a useful method in simulating the motion of a soccer ball and adjusting the design parameters for the optimization design of the flight trajectory.

4.2 Optimization Modeling and Method Implementation

In our team, the optimization modeling of a soccer ball motion is currently being investigated at the research and technology development levels. The technology will be used to improve the initial parameters for shooting a goal. One article has been published to introduce the optimal method and implementation to predict a soccer target through dynamic modeling and simulation (Li et al. 2020). In this Chapter, the detail of the study is introduced for guiding the students and scholars to do further studies.

Y. Li, *Motion Analysis of Soccer Ball*,
SpringerBriefs in Applied Sciences and Technology,
https://doi.org/10.1007/978-981-16-8652-8_4

The optimization design of soccer ball motion involves three factors: (a) design variables—initial parameters design, (b) design constraints—required range, and (c) design objective—expected target. Three factors are the keys in the optimization modeling process and method implementation.

In a soccer multi-body dynamics model, if there are n design variables and each one is expressed by x_i (i = 1, 2, 3, ..., n), then all design variables can be donated by a matrix X as given in Eq. 4.1.

$$X = [x_1, x_2, \ldots, x_n]^T \tag{4.1}$$

If the objective function is donated by f(X), which is related to the design variables x_i (i = 1, 2, 3, ..., n), then objective function f(X) can be expressed as a function of the design variables X as indicated in Eq. 4.2.

$$f(X) = f(x_1, x_2, \ldots, x_n) \tag{4.2}$$

Next, selecting the minimum objective function f(X), the optimization model is derived by Eq. 4.3.

$$min\ f(X) = min\ f(x_1, x_2, \ldots, x_n) \tag{4.3}$$

Furthermore, assuming to require m design constraints to apply for the model and each one is expressed by g_u (X) (u = 1, 2, 3, ..., m), then the general Equation of the design constraints are expressed as Eq. 4.4.

$$g_u(X) = g_u(x_1, x_2, \ldots, x_n) \geq 0 \quad (u = 1, 2, \ldots, m) \tag{4.4}$$

Note, g_u (X) is associated with design variables x_i (i = 1, 2, 3, ..., n). Integration of Eqs. 4.1–4.4 provides a basic optimization modeling to implement design processes repeatedly for finding the design parameters within satisfying design constraints.

The general procedure has been summarized in Fig. 4.1, which presents a flow chart from the dynamics modeling through dynamic simulation and analysis to design optimization for achieving a soccer ball shooting at target. Our previous studies (Li et al. 2020) have laid a foundation for the programming the procedure. However, the program missed detailed steps. This procedure adds more steps, such as dynamic moment calculation as given in simulation section in Fig. 4.1. The details are summarized as following steps.

Step 1: A case study objective is planned, which can be shooting at a goal or a target.
Step 2: Virtual prototype of the field-goal-ball is created, where initial position is defined, and parameters are configurated. And then, the equations of motion are loaded on the multi-body dynamics system.

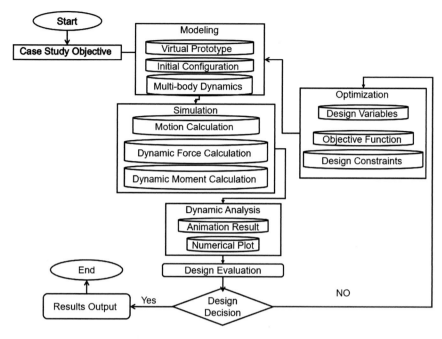

Fig. 4.1 A flowchart of modeling, simulation, and optimization of predicting soccer ball shooting at target

Step 3: The algorithm goes to the simulation stage. The motion parameters are computed. The dynamic forces and dynamic moments are calculated for each instantaneous position of the ball.

Step 4: The simulation results are output, which include the animation to visualize the motion and force/moment and the numerical diagram to plot all data parameters.

Step 5: The design results are analyzed, compared, and evaluated.

Step 6: If design is not satisfied, the program turns on the optimization operation to adjust the design variables. In this book, case study requires minimize the objective function within the constraint ranges.

Step 7: The parameterization functions are embedded in the modeling, which can enforce design relationships to fully parameterize model. The advantage of parameterization is to change initial configuration of model automatically.

Step 8: The design variables are examined to estimate if they are set within the constraint ranges.

Step 9: All design variables are brought into a mathematical expression by the objective function to evaluate results.

Step 10: Checking if all constraints are defined to ensure the optimized design variables within overall limits.

Step 11: Improving the model design through the iterative design by modifying the initial parameters to meet the objective requirements.

Step 12: The optimization operation finds the best values of the design variables.
Step 13: Optimized design results are output, and program has been ended.

4.3   A Case Study of a Soccer Ball Shooting at Target

In Chap. 3, a case study is given for the dynamic simulation and analysis of a soccer ball projectile motion. In this Chapter, continuing the case study, a soccer ball shooting at target is simulated and its flight trajectory is predicted. In our early study (Li et al. 2020), the optimal method has been proposed and a case study has been presented to predict a soccer target throughout dynamics modeling and optimizing design parameters. In this Chapter, the work has extended to wide-field and more detailed aspects have been described. The works focus on (a) the initial configuration of field-goal-ball, (b) the virtual prototype modeling of field-goal-ball, (c) the parameter design of ball's velocity and position, (d) the multi-body dynamics of system motion, (e) the contact modeling of ball-body, and (f) the dynamic simulation and results analysis. Tables 3.1 and 3.2 summarize the general information used in this case study, which includes the geometric parameters, physical properties, and initial conditions. The details are indicated in the following paragraphs.

4.3.1   Initial Configuration and Dynamics Modeling

Figure 4.2 shows a multi-body dynamics model to simulate a soccer ball shooting at goal in a virtual field environment. Figure 4.2a shows the initial configuration of soccer field-goal-ball. Table 4.1 lists the geometric parameters and physical properties used for modeling. The virtual prototypes involved in the virtual environment are the solid geometric bodies: (1) a virtual soccer field in the length LF = 120 m and width WF = 90 m; (2) a virtual goal area in length LA = 18.32 m and width WA = 5.5 m; (3) a virtual soccer goal in the width WG = 7.32 m and height HG = 2.44 m; (4) a virtual soccer ball in the diameter d = 0.2286 m and the mass m = 0.43 kg.

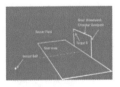

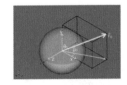

(a) Initial configuration of soccer field-goal-ball   (b) Soccer ball flight trajectory   (c) Free-body diagram of soccer ball with an initial velocity

Fig. 4.2 Multi-body dynamics modeling and simulation of a soccer ball shooting at goal in a virtual environment

Table 4.1 Geometric parameters and physical properties

Geometric parameters								Physical properties	
Field		Goal area		Goal		Ball		Environment	
Length	Width	Length	Width	Width	Height	Diameter	Mass	Air density	Drag coefficient
LF (m)	WF (m)	LA (m)	WA (m)	WG (m)	HG(m)	d (m)	m (kg)	ρ (kg/m$^{3)}$	C$_d$ (kg/m)
120	90	28.32	5.5	7.32	2.44	0.2286	0.43	1.205	0.2

Table 4.2 Initial condition

Initial parameters		
Initial velocity	Initial projectile angle	Initial orientation angle
v_0 (m/s)	θ_0 (°)	β_0 (°)
12	60	30

The initial conditions are defined in Table 4.2: a soccer ball is initially at rest on the field and is launched with an initial velocity $v_0 = 12\ m/s$ at an initial projectile angle $\theta_0 = 60°$, and an initial orientation angle $\beta_0 = 30°$ (Li and Li 2019).

Figure 4.2b presents a flight trajectory of the soccer ball shooting at goal. A local Cartesian coordinate system o(x, y, z) is attached to the field at the center of the ball, where original point o is 10 m away from the goal and 3.66 away from the central line of the goal area, x–z plane attached on the field with x-axis along the length direction, z-axis along the width direction, and y-axis is perpendicular to the field.

Figure 4.3c indicates a diagram with initial velocity vector $\vec{v}_0$, initial projectile angle θ_0, initial orientation angle β_0. Initial velocity vector $\vec{v}_0$ can be written as three scalar components v_{x0}, v_{y0}, and v_{z0} in x-y-z directions.

$$v_{x0} = v_0 \cos \theta_0 \cos \beta_0 \tag{4.5a}$$

$$v_{y0} = v_0 \sin \theta_0 \tag{4.5b}$$

$$v_{z0} = -v_0 \cos \theta_0 \sin \beta_0 \tag{4.5c}$$

The dynamics model of the soccer ball is established under the four assumptions (Li and Li 2019): (1) the air surrounding the ball to be homogenous, (2) ignoring drag moment, (3) ignoring the Magnus effect on the ball or the ball flying spin, (4) ignoring the gyroscopic moment of the ball, and (5) neglecting acceleration of Coriolis.

Figure 4.3 shows a free-body diagram of the ball in an instantaneous position. The forces applied on the ball include the gravity $\overrightarrow{G}$ and aerodynamic drag $\overrightarrow{F}_d$. The Equations of Motion are governed by Newton Equations (see Chap. 3).

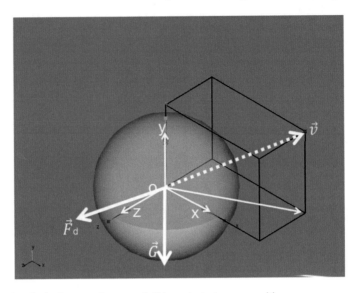

Fig. 4.3 Free-body diagram of a soccer ball in an instantaneous position

$$\sum \vec{N} = m\vec{a} \tag{4.6}$$

where $\sum \vec{N}$ is the vector summation of the general external forces; m is the mass of the ball; $\vec{a}$ is the acceleration vector of the ball. The forces applied to the ball are given in the following. Several equations have been derived in Chap. 3 and are repeated in this Chapter for the sake of completeness.

Gravity: gravity $\vec{G}$ acts to the ball in -y-direction and is calculated by Eq. 4.7.

$$\vec{G} = -mg \tag{4.7}$$

where g = 9.8 m/s^2 is the acceleration due to gravity.

Aerodynamic Drag: the aerodynamic drag $\vec{F}_d$ acts to the ball in the direction opposite to its flying velocity $\vec{v}$ and is calculated by Eq. 4.8 (Bray and Kerwin 2003; Asai et al. 2007, 2013; Goff and Carré 2009).

$$\vec{F}_d = -\frac{1}{2}\rho C_d A \vec{v}\lfloor \vec{v}\rfloor = -K_d \vec{v}|\vec{v}| \tag{4.8}$$

where ρ is the density of the air, $\rho = 1.205$ kg/m^3 for the environment temperature to be 20^0 C; A is the cross-sectional area of the ball; and C_d is the drag coefficient. The drag coefficient depends on the boundary conditions, such as the surface roughness

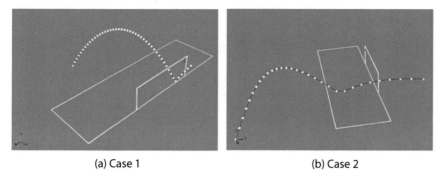

(a) Case 1   (b) Case 2

Fig. 4.4 Two cases of a soccer ball impacting on field and then shooting at goal

of a ball, as well as the laminar flow or turbulence of the atmospheric layer. Smith et al. (1999) suggested $C_d = 0.2$. If d denotes the diameter of the ball, then,

$$K_d = \frac{1}{8}\rho C_d \pi d^2 \tag{4.9}$$

Then, the aerodynamic drag $\vec{F}_d$ is broken down into three component forces, F_{dx}, F_{dy} and F_{dz}, in x-y-z directions, respectively. Therefore, vector Eq. 4.8 can be written as three sets of scalar Eqs. 4.10a–c.

$$F_{dx} = -K_d \dot{x} |\dot{x}| \tag{4.10a}$$

$$F_{dy} = -K_d \dot{y} |\dot{y}| \tag{4.10b}$$

$$F_{dz} = -K_d \dot{z} |\dot{z}| \tag{4.10c}$$

Contact Force: if the ball hits on the field (see Fig. 4.4) or on the goal (see Fig. 4.5), the ball-body contact model can be either the ball-field or ball-woodwork. In our previous research (Li et al. 2020), two cases have been described as demonstrated in Figs. 4.4a and 4.5a. Here, two more cases are depicted as given in Figs. 4.4b and 4.5b. Figure 4.6a and b shows the configurations of the ball-field and ball-woodwork contact with the mass-dumping-stiffness properties. The equation of the contact force is derived by employing Newtown's Second Law and Hooke's law.

$$\vec{F}_n = \vec{F}_c + \vec{F}_k = c\vec{v} + k\vec{\delta} \tag{4.11}$$

where $\vec{F}_c$ is the damping force vector, $\vec{F}_k$ is the elastic force vector, c is the contact damping coefficient ($c = 3.8 \times 10^5$ N/m), $\vec{v}$ is the velocity vector, k is the contact

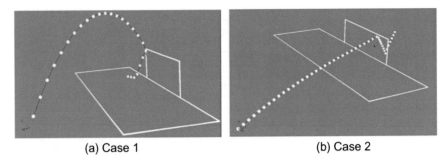

(a) Case 1   (b) Case 2

Fig. 4.5 Two cases of a soccer ball hitting on woodwork

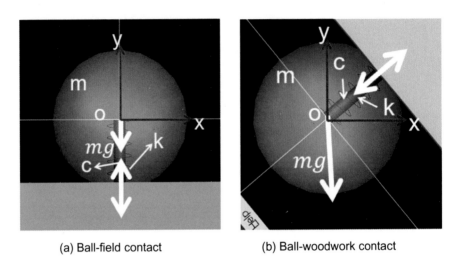

(a) Ball-field contact   (b) Ball-woodwork contact

Fig. 4.6 Two diagrams of ball-body contact model

stiffness (k = 97.2 N.s/m), $\vec{\delta}$ is the displacement vector. Note the contact damping coefficient and contact stiffness cannot find in material property handbooks. The data given here are calculated results. The contact force $\vec{F}_n$ is separated into three scalar components F_{nx}, F_{ny}, and F_{nz} in x-y-z directions.

$$F_{nx} = c\dot{x} + kx \tag{4.12a}$$

$$F_{ny} = c\dot{y} + ky \tag{4.12b}$$

$$F_{nz} = c\dot{z} + kz \tag{4.12c}$$

Substituting the gravity $\vec{G}$ and aerodynamic drag $\vec{F}_d$ and contact force $\vec{F}_n$ into Eq. 4.6, it can be written as

$$\sum \vec{N} = m\vec{a} = \vec{G} + \vec{F}_d + e\vec{F}_n \tag{4.13}$$

where, $e = 0$, when $\vec{\delta} > = d/2$, there is no penetration between the ball and body and the contact force is zero; and $e = 1$, when $\vec{\delta} < d/2$, there is penetration between the ball and body and the contact force is larger than zero.

Equation 4.13 can be separated into differential Equations in x-y-z directions. There are three components of motions generated as the ball and body coming into contact in x-y-z directions. The motions then are given as Eqs. 4.14a–c, in x-y-z directions, respectively.

$$m\ddot{x} = -K_d\dot{x}|\dot{x}| + e(c\dot{x} + kx) \tag{4.14a}$$

$$m\ddot{y} = -mg - K_d\dot{y}|\dot{y}| + e(c\dot{y} + ky) \tag{4.14b}$$

$$m\ddot{z} = -K_d\dot{z}|\dot{z}| + e(c\dot{z} + kz) \tag{4.14c}$$

Equations 4.14a–c can be expressed as a matrix Eq. 4.15.

$$\begin{bmatrix} m & & \\ & m & \\ & & m \end{bmatrix} \begin{bmatrix} \ddot{x} \\ \ddot{y} \\ \ddot{z} \end{bmatrix} = \begin{bmatrix} -K_d\dot{x}|\dot{x}| + e(c\dot{x} + kx) \\ -mg - K_d\dot{y}|\dot{y}| + e(c\dot{y} + ky) \\ -K_d\dot{z}|\dot{z}| + e(c\dot{z} + kz) \end{bmatrix} \tag{4.15}$$

This matrix forms a kinematics and dynamics system of 3 equations and 3 unknowns, $\ddot{x}$, $\ddot{y}$, and $\ddot{z}$. The simultaneous constraint method (Gardner 2001) is employed to solve Eq. 4.15.

Figure 4.7 shows the dynamic simulation diagram of this system. There is no input system. This equation may be solved subject to the initial conditions. Equations 4.5a–c will determine its response. This matrix equation is embedded in an ODEs solution system. The velocities ($\dot{x}$, $\dot{y}$, and $\dot{z}$) and displacements (x, y, and z) can be computed by integrating accelerations $\ddot{x}$, $\ddot{y}$, and $\ddot{z}$. Therefore, if initial configuration, parameters design, and dynamics modeling are all given, the position-time history of a ball can be obtained. A case study is presented in next section to simulate the soccer ball shooting at target.

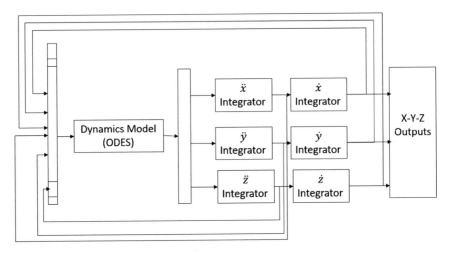

Fig. 4.7 Dynamic simulation diagram of a soccer ball projectile motion

4.3.2   Dynamic Simulation and Results Analysis

The dynamic simulation is performed in a virtual soccer field-goal-ball environment. Figure 4.8 shows dynamic simulation of the soccer ball flight vs time 2.5 s. This case has been inherited from our studies in the first stage (Li et al. 2020). The soccer ball is launched from initial point o(0, 0, 0) with initial velocity $v_0 = 12$ m/s, initial projectile angle $\theta_0 = 60°$, and initial orientation angle $\beta_0 = 30°$. The animation displays the ball traveling trajectory (see Fig. 4.7a) and Fig. 4.7b reveals the ball position at instantaneous point A (x, y, z). It notes that the ball traveling trajectory exhibits a parabolic shape and reaches the goal at point B', which almost touches the field. The target B(10, 2.04, −7) is located at the right-top corner. So, the flight trajectory shows the ball does not get close to target B.

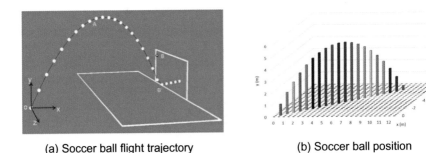

(a) Soccer ball flight trajectory   (b) Soccer ball position

Fig. 4.8 Dynamic simulation of the soccer ball flight vs time 2.5 s with initial velocity $v_0 = 12$ m/s, initial projectile angle $\theta_0 = 60°$, and initial orientation angle $\beta_0 = 30°$

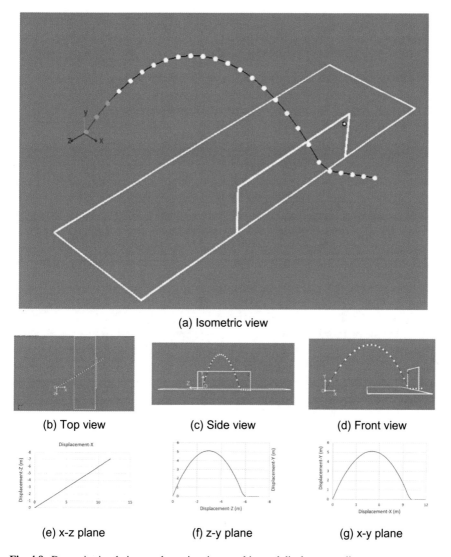

Fig. 4.9 Dynamic simulation results: animation graphics and displacement diagrams

Figure 4.9a–g shows the dynamic simulation results. The animation displays the traveling trajectory in four views: (a) isometric view, (b) top view, (c) side view, and (d) front view. The displacement of the ball is plotted in coordinate system o(x, y, z) with three planes: (e) x-z plane, (f) z-y plane, and (g) x-y plane.

It can find how far away the ball from point B' to point B, such as for x = 10 m, then z = −6m (see Fig. 4.9e); and for x = 10 m, then y = 0 m (see Fig. 4.9g). That means that the ball reaches the goal at point B'(10, 0, −6) and 2.27 m away from target B(10, 2.04, −7). For reaching target B, it is necessary to use an optimization design method to find the best initial parameters, which are the initial velocity, initial projectile angle, and initial orientation angle.

4.4 Design Improvement Through Parameters Optimization

The design improvement is realized to achieve the ball shooting at target B. Three initial parameters (initial velocity, initial projectile angle, and initial orientation angle) have been changed to optimize the traveling trajectory. The problem is stated in this: modify three initial parameters for the ball to reach target B. And then explore the best initial parameters, initial velocity v_0, initial projectile angle θ_0, and initial orientation angle β_0. The solution of the problem has been reported in our publication (Li et al. 2020). In this Section, the solution has been updated and presented in more detail.

Creating Design Variable: there are three independent design variables u_i ($i = 1, 2, 3$) that participate in this case study. They are: (a) an initial velocity $u_1 = v_0$ with a standard value of 12 m/s and the required range between 0 and 30 m/s; (b) an initial projectile angle $u_2 = \theta_0$ with a standard value of 60° and the required range between 0 and 90°; and (c) an initial orientation angle $u_3 = \beta_0$ with a standard value of 30° and the required range between 0 and 180°. If three design variables are assembled into a matrix U, it can be expressed as Eq. 4.16.

$$U = [v_0, \theta_0, \beta_0]^T = [u_1, u_2, u_3]^T \tag{4.16}$$

Optimizing Objective Function: Fig. 4.8 displays that a soccer ball is launched from initial point o(0, 0, 0), flies through the air, and expected to arrive at target B(x_b, y_b, z_b). The distance L is measured from an instantaneous center of the ball A (x, y, z) to the target B(x_b, y_b, z_b). The optimization objective is assigned as the minimum distance (MD) L. If a tolerance is defined, it is acceptable when the objective value has smaller than this tolerance. The objective function f(U) is created in Cartesian coordinate system o(x, y, z) as shown in Eq. 4.17.

$$f(U) = L = \sqrt{(x_b - x)^2 + (y_b - x)^2 + (z_b - x)^2} \tag{4.17}$$

Defining Design Constraints: in this case, the initial position of the ball has been fixed on the field and it cannot be changed. The optimization iterating process can only adjust three design variables u_1, u_2, and u_3 within the required ranges, which are provided in the above Section (Creating Design Variable). In this optimization program, all constraints are set larger or equal to zero. Therefore, the boundary conditions of the three design variables are defined as Eqs. 4.18–4.23 (Li et al. 2020).

$$g_1(U) = u_1 > 0 \tag{4.18}$$

$$g_2(U) = 30 - u_1 \geq 0 \tag{4.19}$$

$$g_3(U) = u_2 \geq 0 \tag{4.20}$$

$$g_4(U) = 60 - u_2 \geq 0 \tag{4.21}$$

$$g_5(U) = u_3 \geq 0 \tag{4.22}$$

$$g_6(U) = 180 - u_3 \geq 0 \tag{4.23}$$

Iterating Optimization Process: this part task is to improve the initial design parameters for meeting the objective requirements. The program of the optimization can automatically find a better path to approach the minimization of the objective function f(U). The iterative operation is processed by varying design variables u_1, u_2, and u_3, checking and keeping u_1 within the range [0, 30 m/s], u_2 within the range [0, 90°], and u_3 within the range [0, 180°]. For each iteration, the optimal program produces three new values u_1, u_2, and u_3. And next, the program repeats the operation process for the new iteration. If the calculated difference of the objective values between the current iteration and the last one is smaller than the specified tolerance 0.04 m or specified one, the optimization operation is completed, and results are output.

Outputting Iteration Results: up to this step, the problem is near solved. The iteration times depend on the tolerance. For each iteration, the simulating results are automatically examined to see if the data is available to output. This operation process of iteration has been used by Li et al. (2020). It is also used in this solution to further illustrate the method. Figure 4.10 shows the variations of the measured distance L vs time t = 2.5 s for nine iterations. The results are plotted as distance-time curves, which implies that nine curves change in the similar tendencies. Looking at each curve, distance L decreases within the time frame {0–1.9 s}, reaches the minimum value at time 1.9 s, and then increases within the time frame {1.9–2.5 s}. It can find at 1.9 s, the distance L has minimum value, the ball arrives at goal with the traveling time of 1.9 s. In Fig. 4.10, there is a zooming in the area around time 1.9 s. The distribution of MD L vs iteration is exhibited clearly to further find details. The final result indicates that it needs to take nine iterations to achieve the outcome. Thus, the MD L experiences the greatest value of 1.866 m at the first iteration and drops down to the smallest value of 0.037 m at the ninth iteration. It is observed that from the iteration 1 to iteration 2, the MD L generates a big gap. However, for the other eight iterations, the MD L goes very smoothly with time and then at time 1.9 s it

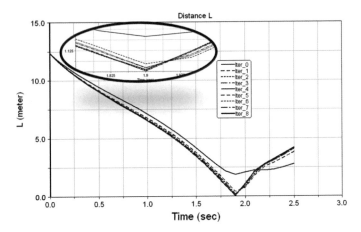

Fig. 4.10 Distance-time curves of soccer ball flight with nine iterations from 0 to 2.5 s

varies from 0.414 down to 0.037 m. Finally, the ball moves closest to point B while operation finishes the ninth iteration.

Determining Optimized Parameters: The program reports the optimal design results and generates a report for nine iterations in Table 4.3. The initial values and final values of MD L and design parameters nine iterations for nine iterations have been listed. The design objectives include initial and final MD L values. Also, the design variables list the initial and final values of initial velocity u_1, initial projectile angle u_2, and initial orientation angle u_3. The comparison of the original design parameters (standard parameters) and the ninth iteration results (optimized parameters) are concluded in Table 4.3. It can be seen that the design objective MD L drops 98%, the initial velocity rises 7.03%, the initial projectile angle loses 3.82%, and the initial orientation angle obtains 15.3%. The iteration results in the last step present that the optimized design objective is 0.037 m within the optimal combination of three design variables: (a) initial velocity $u_1 = v_0 = 12.84$ m/s, (b) initial projectile angle $u_2 = \theta_0 = 57.71°$, and (c) initial orientation angle $u_3 = \beta_0 = 34.58°$.

Optimization Results and Analysis: Fig. 4.11 shows dynamic simulation of the soccer ball flight vs time 2.5 s. The soccer ball is launched from initial point o(0, 0, 0) with initial velocity $v_0 = 12.85$ m/s, initial projectile angle $\theta_0 = 57.71°$, and initial orientation angle $\beta_0 = 34.58°$. Figure 4.11a displays the ball traveling trajectory and Fig. 4.11b reveals the ball position at instantaneous point A (x, y, z). It shows that the flight path exhibits a parabolic shape. Comparing the animation result in Fig. 4.8, the actual target B' moves from the right-bottom corner (see Fig. 4.8a) to the right-top corner (see Fig. 4.11a). The optimized result shows the ball gets close to the expected target B located in x = 10, y = 2.04, and z = −7.

Figure 4.12a–g shows the dynamic simulation results. The animation displays the flight trajectory in four views: (a) isometric view, (b) top view, (c) side view, and (d) front view. The displacement of the ball is plotted in coordinate system o(x, y, z) with

Table 4.3 The optimized design results of nine iterations

Design objectives	Minimum of measurement distance L			
	Units: meter			
	Initial value:	1.86566		
	Final value:	0.0373784 (−98%)		
Design variables	(u1) Initial velocity			
	Units: meter/s			
	Initial value:	12		
	Final value:	12.8436 (+7.03%)		
	(u2) Initial projectile angle			
	Units: deg			
	Initial value:	60		
	Final value:	57.7052 (−3.82%)		
	(u3) Initial orientation angle			
	Units: deg			
	Initial value:	30		
	Final value:	34.5761 (+15.3%)		
Iteration results	Iteration	L	u1	u2	u3
	0	1.8657	12.000	60.000	30.000
	1	0.41454	12.872	58.558	33.188
	2	0.21742	12.943	58.043	34.480
	3	0.20106	12.955	57.843	34.324
	4	0.11674	12.891	57.491	34.451
	5	0.079345	12.844	57.498	34.231
	6	0.048698	12.818	57.601	34.327
	7	0.028697	12.827	57.710	34.301
	8	0.037378	12.844	57.705	34.576

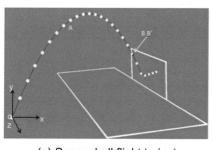

(a) Soccer ball flight trajectory

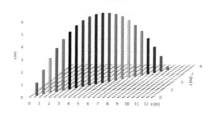

(b) Soccer ball position

Fig. 4.11 Dynamic simulation of the soccer ball flight versus time 2.5 s with initial velocity $v_0 = 12.85$ m/s, initial projectile angle $\theta_0 = 57.71°$, and initial orientation angle $\beta_0 = 34.58°$

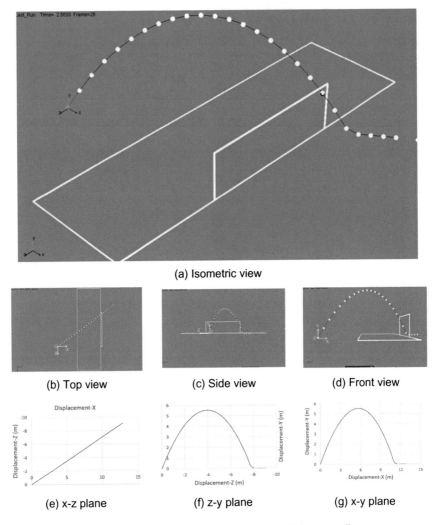

(a) Isometric view

(b) Top view   (c) Side view   (d) Front view

(e) x-z plane   (f) z-y plane   (g) x-y plane

Fig. 4.12 Dynamic simulation results: animation graphics and displacement diagrams

three planes: (e) x-z plane, (f) z-y plane, and (g) x-y plane. The accurate position of point B' can be found in Fig. 4.12. For example, Fig. 4.12e shows that for x = 10 m, then z = −7.03 m and Fig. 4.12g shows for x = 10 m, then y = 2.05 m. Therefore, it can say that the ball leaves the player's foot at $v_0 = 12.85$ m/s from about 10 m away from the goal. The ball flies from point (0, 0, 0) with initial projectile angle θ_0 = 57.71°, and initial orientation angle $\beta_0 = 34.58°$. The ball rises 2.04 m height and moves laterally 7 m and dips into the corner of the goal. The target point B'(x = 10, y = 2.05, and z = −7.03) stays 0.0374 m away from the desired target B(x = 10, y = 2.04, and z = −7).

General Summary: This chapter indicates the actual application of the proposed method of the optimization design for the planning flight trajectory of the soccer ball shooting at goal. The procedure operation is expected to best meet the performance parameters while satisfying the design constraints. The running process is designed considering the design requirements and tolerance. For the case study demonstrated in this Chapter, the procedure of the optimization program involves three steps.

1. Creating the objective function considering achieving the measuring distance L to be minimized.
2. Determining the three design variables which can be changed within the required ranges.
3. Defining the constraint ranges and checking if the results are satisfying.

The optimized results are output as diagrams and tables, animation graphics. The work can be described in five aspects.

1. Presenting the design variables, which are obtained in final iteration and are optimized results.
2. Stating the improved design parameters and performing simulation again.
3. Visualizing the optimized trajectory and comparing it to the original design.
4. Plotting the data in diagrams and analyzing the data.
5. Declaring the combination of the most optimal parameters to achieve a successful outcome.

Therefore, the solution for the problem can be stated as: when the MD difference of 0.0374 m and the tolerance range of 0.04 m are specified, the soccer ball can reach at target B with initial velocity $v_0 = 12.85$ m/s, initial projectile angle $\theta_0 = 57.71°$, and initial orientation angle $\beta_0 = 34.58°$.

References

Asai T, Seo K, Kobayashi O, Sakashita R (2007) Fundamental aerodynamics of the soccer ball. Sports Eng 10:101–109

Asai T, Seo K (2013) Aerodynamic drag of modern soccer balls. Springerplus 2:171

Bray K, Kerwin DG (2003) Modelling the flight of a soccer ball in a direct free kick. J Sport Sci 21:75–85

Goff JE, Carré MJ (2009) Trajectory analysis of a soccer ball. Am J Phys 77:1020–1027

Gardner J (2001) Simulations of machines using Matlab and Simulink, 1st edn. University of California, Brooks

Li Y, Li Q (2019) Soccer ball spatial kinematics and dynamics simulation for efficient sports analysis. Asian J Adv Res Rep 7(4):1–18. ISSN: 2582-3248

Li Y, JX M, Li Q (2020) Predicting soccer ball target through dynamic simulation. J Eng Res Rep 12(4):6–18

Smith MR, Hilton DK, Van Sciver SW (1999) Observed drag crisis on a sphere in flowing He I and He II. Phys Fluids 11:751–753

Chapter 5
Modeling and Simulation of Soccer Ball Free Kick

5.1 Overview

In this Chapter, the dynamics modeling, and virtual simulation of soccer ball free kick are investigated. A three body dynamics model is created based on the Newton second law and Hooke's law. The model includes gravity, aerodynamic drag, Magnus force, and contact force. The motion of soccer ball is established as the time-dependent ordinary differential equations (ODEs). The virtual prototype technology is introduced and applied for the dynamic simulation of soccer ball free kicks. Some typical results are obtained.

5.2 Initial Configuration

Figure 5.1 shows a multi-body dynamics model of a soccer ball to simulate a free kick in a virtual field environment. The ball starts from point o and travels through the air to a selected target B at the goal. The model consists of three major components, which are one soccer field, one goal, and one soccer ball. Three components are made of different materials and are regarded as a multi-body system (Li et al. 2020). They are related by some restrictions. The field is fixed in the ground and provides a static platform for holding the ball and goal. The goal area is the small box inside the field. The goal is structured by three bars (one crossbar and two goalposts) and mounted on the field by the feet of two goalposts. Their geometric parameters are (1) the soccer field: length 120 m and width 90 m; (2) the goal area: length 18.32 m and width 5.5 m; (3) the soccer goal: width 7.32 m and height 2.44 m; and (4) the soccer ball: diameter d = 0.2286 m and mass m = 0.43 kg. Table 5.1 summarizes the geometric parameters of field-goal-ball.

© The Author(s), under exclusive license to Springer Nature Singapore Pte Ltd. 2022   67
Y. Li, *Motion Analysis of Soccer Ball*,
SpringerBriefs in Applied Sciences and Technology,
https://doi.org/10.1007/978-981-16-8652-8_5

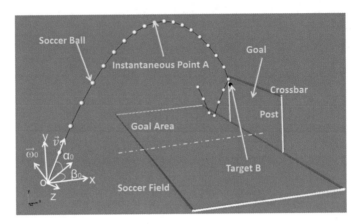

Fig. 5.1 Multi-body dynamics modeling of a soccer field-goal-ball in a virtual environment

Table 5.1 The geometric parameters of field-goal-ball

Field		Goal area		Goal		Ball	
Length	Width	Length	Width	Width	Height	Diameter	Mass
LF (m)	WF (m)	LA (m)	WA (m)	WG (m)	HG(m)	d (m)	m (kg)
120	90	28.32	5.5	7.32	2.44	0.2286	0.43

A global Cartesian coordinate system O(X, Y, Z) is attached to the field at its center line and 10 m away from the goal and 3.66 away from the central line of the goal area, where X-Z plane is attached on the field with X-axis along the length direction, Z-axis along the width direction, and Y-axis is perpendicular to the field. A local Cartesian coordinate system o(x, y, z) is attached to the field at the ball center in the initial position. The direction of the x-axis is normal toward the goal. The y-axis is perpendicular to the x-axis and normal out of the field. The z-axis is normal to the x-y plane, and its positive direction is out of the x-y plane, which is determined by the right-hand rule.

In initial position, the ball lies on the field and the up-down moving is restricted. Therefore, the ball has five degrees of freedom and only the translation along with y-direction is restricted. However, other five movements are free. So, it can translate along the x and z directions and rotate around the x, y, and z directions. The motion of the ball comprises four distinct phases: (i) the launching at the initial position with an initial velocity and angular velocity in a selected direction (projectile angle and orientation angle), (ii) the spinning around its central axis and the going-up with deceleration and then going-down with acceleration, (iii) the shooting at an expected target, and (iv) the landing on the field.

Figure 5.1 shows that the ball is launched at an initial position (x_0, y_0, z_0) with an initial vector $\overrightarrow{v_0}$, initial projectile angle θ_0 and initial orientation angle β_0. The initial

Fig. 5.2 Free-body diagram of the ball with an initial velocity and angular velocity

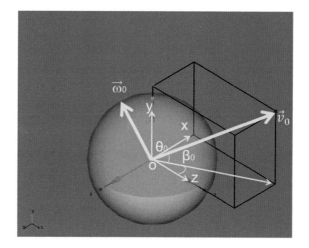

angular velocity $\vec{\omega}_0$ applied on the ball is normal to $\vec{v_0}$. It can be calculated by $\vec{\omega}_0$ $= 3 \, s|\vec{v_0}|/r^2$, where s is the vertical distance measured from the center of the ball to the kick force vector; and r is the radius of the ball.

Figure 5.2 indicates a diagram with initial velocity vector $\vec{v_0}$, initial angular velocity vector $\vec{\omega}_0$, initial projectile angle θ_0, and initial orientation angle β_0. Initial velocity vector $\vec{v_0}$ can be written in three scalar components v_{x0}, v_{y0}, and v_{z0} in x-y-z directions as shown in Eqs. 5.1a–c.

$$v_{x0} = v_0 \cos \theta_0 \cos \beta_0 \tag{5.1a}$$

$$v_{y0} = v_0 \sin \theta_0 \tag{5.1b}$$

$$v_{z0} = -v_0 \cos \theta_0 \sin \beta_0 \tag{5.1c}$$

Similarly, $\vec{\omega}_0$ can be separated into three scalar components ω_{x0}, ω_{y0}, and ω_{z0} in x-y-z directions as shown in Eqs. 5.2a–c.

$$\omega_{x0} = -\omega_0 \sin \theta_0 \cos \beta_0 \tag{5.2a}$$

$$\omega_{y0} = \omega_0 \cos \theta_0 \tag{5.2b}$$

$$\omega_{z0} = \omega_0 \sin \theta_0 \sin \beta_0 \tag{5.2c}$$

5.3 Dynamics Modeling and Dynamic Simulation

5.3.1 Dynamics Modeling

Figure 5.3 shows a free-body diagram of the ball in an instantaneous position. The force applied on the ball can be the summation of the general external forces (such as the aerodynamics drag, Magnus force, and gyroscopic moment) and the inertia forces (such as gravity and acceleration of Coriolis). In this Chapter, the model is created by assuming air resistance is homogenous and ignoring gyroscopic moment and acceleration of Coriolis. Therefore, the forces applied on the ball include the gravity $\vec{G}$, aerodynamic drag $\vec{F_d}$, Magnus force $\vec{F_m}$. The dynamics equations are governed by Newton–Euler Equations (Synge and Griffith 1959).

$$\sum \vec{N} = m\vec{a} \tag{5.3a}$$

$$\sum \vec{M} = I\vec{\varepsilon} = 0 \tag{5.3b}$$

where $\sum \vec{N}$ is the vector summation of the general external forces; m is the mass of the ball; $\vec{a}$ is the acceleration vector of the ball; $\sum \vec{M}$ is the vector summation of the general external moments; I is the moment of inertia about the center of mass of the ball; $\vec{\varepsilon}$ is the angular acceleration vector of the ball. The motions and forces applied to the ball are established as following Equations.

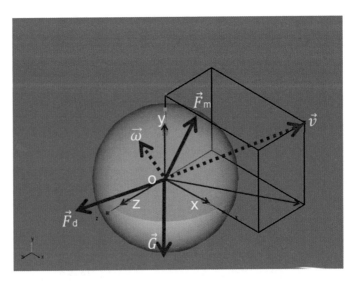

Fig. 5.3 Free-body diagram of a soccer ball in an instantaneous position

Gravity: gravity $\vec{G}$ acts to the ball in -y-direction and is calculated by Eq. 5.4.

$$\vec{G} = -mg \tag{5.4}$$

where $g = 9.8$ m/s^2 is the gravitational acceleration.

Aerodynamic Drag: the aerodynamic drag force $\vec{F_d}$ acts to the ball in the direction opposite to the ball flying velocity $\vec{v}$ and is calculated by Eq. 5.5 (Bray and Kerwin 2003; Asai et al. 2007, 2013; Goff and Carré 2009).

$$\vec{F_d} = -\frac{1}{2}\rho C_d A \vec{v} \lfloor \vec{v} \rfloor = -K_d \vec{v} |\vec{v}| \tag{5.5}$$

where ρ is the density of the air, $\rho = 1.205$ kg/m^3 for the environment temperature to be 20 °C; A is the cross-sectional area of the ball; and C_d is the drag coefficient. The drag coefficient depends on the boundary conditions, such as the surface roughness of a ball, as well as the laminar flow or turbulence of the atmospheric layer. It is an important parameter to describe the behavior of a spinning soccer ball during flying through the air and can be calculated based on the investigations of Asai et al. (2007) and Smith et al. (1999). In this Chapter, $C_d = 0.2$ is selected for all case studies. If d denotes the diameter of the ball, then,

$$K_d = \frac{1}{8}\rho C_d \pi d^2 \tag{5.6}$$

Then, the aerodynamic drag force $\vec{F}_d$ is broken down into three component forces, F_{dx}, F_{dy}, and F_{dz}, in x-y-z directions, respectively. Therefore, vector Eq. 5.5 can be written as three sets of scalar Eqs. 5.7a–c.

$$F_{dx} = -K_d \dot{x} |\dot{x}| \tag{5.7a}$$

$$F_{dy} = -K_d \dot{y} |\dot{y}| \tag{5.7b}$$

$$F_{dz} = -K_d \dot{z} |\dot{z}| \tag{5.7c}$$

Magnus Force: The Magnus force is caused by Magnus effect and occurs when the ball is spinning while flying through the air. The Magnus effect is the tendency of a spinning, translating ball to be deflected laterally, that is, in a direction perpendicular to both its spin axis and its direction of motion (Bush 2013). For a ball in flight with velocity $\vec{v}$ and spinning with angular velocity $\vec{\omega}$, in addition to drag, there is a Magnus force $\vec{F_m}$ acts to the ball in the direction of the $\vec{\omega} \times \vec{v}$. It is perpendicular to both angular velocity $\vec{\omega}$ and velocity $\vec{v}$, and is calculated by Eq. 5.8.

$$\vec{F}_m = \frac{1}{2}\rho C_m A \vec{v}^2 \frac{\vec{\omega} \times \vec{v}}{\lfloor \vec{\omega} \times \vec{v} \rfloor} = K_m \vec{v}^2 \frac{\vec{\omega} \times \vec{v}}{\lfloor \vec{\omega} \times \vec{v} \rfloor} \qquad (5.8)$$

where C_m is the Magnus coefficient. For a soccer ball flight with spinning, it is also an important parameter to describe the behavior of a spinning soccer ball during flying through the air and can be calculated based on the investigations of Asai et al. (2007) and Goff and Carre (2010). In this chapter, $C_m = 0.2$ is assigned for all case studies. Area A and velocity $\vec{v}$ are the same values used for aerodynamic drag calculation. Factor K_m is expressed as

$$K_m = \frac{1}{8}\rho C_m \pi d^2 \qquad (5.9)$$

Then, the Magnus force $\vec{F}_m$ is separated into three components, F_{mx}, F_{my}, and F_{mz}, in x-y-z directions, respectively. Therefore, vector Eq. 5.8 can be written as three sets of scalar Eqs. 5.10a–c.

$$F_{mx} = K_m \dot{z}^2 \frac{\omega_y\, v_z}{\lfloor \omega_y\, v_z \rfloor} - K_m \dot{y}^2 \frac{\omega_z\, v_y}{\lfloor \omega_y\, v_y \rfloor} \qquad (5.10a)$$

$$F_{my} = K_m \dot{x}^2 \frac{\omega_z\, v_x}{\lfloor \omega_z\, v_x \rfloor} - K_m \dot{z}^2 \frac{\omega_x\, v_z}{\lfloor \omega_x\, v_z \rfloor} \qquad (5.10b)$$

$$F_{mz} = K_m \dot{y}^2 \frac{\omega_x\, v_y}{\lfloor \omega_x\, v_y \rfloor} - K_m \dot{x}^2 \frac{\omega_y\, v_x}{\lfloor \omega_y\, v_x \rfloor} \qquad (5.10c)$$

Contact Force: when a ball contacts with a body such as the field, goal crossbar, or goalpost (see Fig. 5.4), a contact force $\vec{F}_n$ is generated between the ball and the body. Its direction is normal to the contact surfaces at contact point. Figure 5.4 shows a model of the ball-body contact with the mass-dumping-stiffness properties. The equation of the contact force is derived by employing Newtown's Second Law and Hooke's law.

$$\vec{F}_n = \vec{F}_c + \vec{F}_k = c\vec{v} + k\vec{\delta} \qquad (5.11)$$

where $\vec{F}_c$ is the damping force vector, $\vec{F}_k$ is the elastic force vector, c is the contact damping coefficient (c = 3.8 × 10^5 N/m), $\vec{v}$ is the velocity vector, k is the contact stiffness (k = 97.2 N.s/m), $\vec{\delta}$ is the displacement vector. Note the contact damping coefficient and contact stiffness cannot find in material property handbooks. The data given here are calculated results. The contact force $\vec{F}_n$ is separated into three scalar components F_{nx}, F_{ny}, and F_{nz} in x-y-z directions as given in Eqs. 5.12a–c.

$$F_{nx} = c\dot{x} + kx \qquad (5.12a)$$

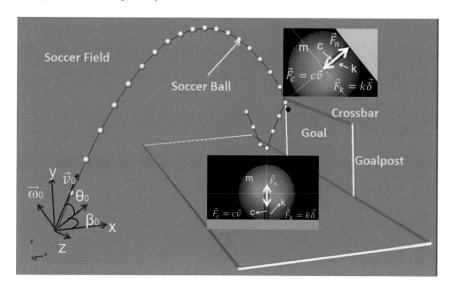

Fig. 5.4 Contact force between the ball and field or the ball and goal

$$F_{ny} = c\dot{y} + ky \tag{5.12b}$$

$$F_{nz} = c\dot{z} + kz \tag{5.12c}$$

So far, Eq. 5.3a can be written as

$$\sum \vec{N} = m\vec{a} = \vec{G} + \vec{F}_d + \vec{F}_m + e\vec{F}_n \tag{5.13}$$

where, $e = 0$, when $\vec{\delta} >= d/2$, there is no penetration between the ball and body and the contact force is zero; and $e = 1$, when $\vec{\delta} < d/2$, there is penetration between the ball and body and the contact force is larger than zero.

Equation 5.13 can be separated into three differential Equations in x-y-z directions.

$$m\ddot{x} = -K_d\dot{x}|\dot{x}| + K_m\dot{z}^2 \frac{\omega_y v_z}{\lfloor \omega_y v_z \rfloor} - K_m\dot{y}^2 \frac{\omega_z v_y}{\lfloor \omega_y v_y \rfloor} + e(c\dot{x} + kx) \tag{5.14a}$$

$$m\ddot{y} = -mg - K_d\dot{y}|\dot{y}| + K_m\dot{x}^2 \frac{\omega_z v_x}{\lfloor \omega_z v_x \rfloor} - K_m\dot{z}^2 \frac{\omega_x v_z}{\lfloor \omega_x v_z \rfloor} + e(c\dot{y} + ky) \tag{5.14b}$$

$$m\ddot{z} = -K_d\dot{z}|\dot{z}| + K_m\dot{y}^2 \frac{\omega_x v_y}{\lfloor \omega_x v_y \rfloor} - K_m\dot{x}^2 \frac{\omega_y v_x}{\lfloor \omega_y v_x \rfloor} + e(c\dot{z} + kz) \tag{5.14c}$$

Assembling Eqs. 5.14a–c, 5.7a–c, 5.10a–c, and 5.12a–c in matrix form, the general dynamics model is represented as Eq. 5.15.

$$
\begin{bmatrix}
m & & & & & & & & & & & \\
& m & & & & & & & & & & \\
& & m & & & & & & & & & \\
& & & 1 & & & & & & & & \\
& & & & 1 & & & & & & & \\
& & & & & 1 & & & & & & \\
& & & & & & 1 & & & & & \\
& & & & & & & 1 & & & & \\
& & & & & & & & 1 & & & \\
& & & & & & & & & 1 & & \\
& & & & & & & & & & 1 & \\
& & & & & & & & & & & 1
\end{bmatrix}
\begin{bmatrix}
\ddot{x} \\
\ddot{y} \\
\ddot{z} \\
F_{dx} \\
F_{dy} \\
F_{dz} \\
F_{mx} \\
F_{my} \\
F_{mz} \\
F_{nx} \\
F_{ny} \\
F_{nz}
\end{bmatrix}
$$

$$
=
\begin{bmatrix}
-K_d \dot{x}|\dot{x}| + K_m \dot{z}^2 \frac{\omega_y\, v_z}{\lfloor \omega_y\, v_z \rfloor} - K_m \dot{y}^2 \frac{\omega_z\, v_y}{\lfloor \omega_y\, v_y \rfloor} + e(c\dot{x} + kx) \\
-mg - K_d \dot{y}|\dot{y}| + K_m \dot{x}^2 \frac{\omega_z\, v_x}{\lfloor \omega_z\, v_x \rfloor} - K_m \dot{z}^2 \frac{\omega_x\, v_z}{\lfloor \omega_x\, v_z \rfloor} + e(c\dot{y} + ky) \\
-K_d \dot{z}|\dot{z}| + K_m \dot{y}^2 \frac{\omega_x\, v_y}{\lfloor \omega_x\, v_y \rfloor} - K_m \dot{x}^2 \frac{\omega_y\, v_x}{\lfloor \omega_y\, v_x \rfloor} + e(c\dot{z} + kz) \\
-K_d \dot{x}|\dot{x}| \\
-K_d \dot{y}|\dot{y}| \\
-K_d \dot{z}|\dot{z}| \\
K_m \dot{z}^2 \frac{\omega_y\, v_z}{\lfloor \omega_y\, v_z \rfloor} - K_m \dot{y}^2 \frac{\omega_z\, v_y}{\lfloor \omega_y\, v_y \rfloor} \\
K_m \dot{x}^2 \frac{\omega_z\, v_x}{\lfloor \omega_z\, v_x \rfloor} - K_m \dot{z}^2 \frac{\omega_x\, v_z}{\lfloor \omega_x\, v_z \rfloor} \\
K_m \dot{y}^2 \frac{\omega_x\, v_y}{\lfloor \omega_x\, v_y \rfloor} - K_m \dot{x}^2 \frac{\omega_y\, v_x}{\lfloor \omega_y\, v_x \rfloor} \\
c\dot{x} + kx \\
c\dot{y} + ky \\
c\dot{z} + kz
\end{bmatrix}
\tag{5.15}
$$

This matrix forms a kinematics and dynamics system of 12 equations and 12 unknowns, $\ddot{x}$, $\ddot{y}$, $\ddot{z}$, F_{dx}, F_{dy}, F_{dz}, F_{mx}, F_{my}, F_{mz}, F_{nx}, F_{ny}, $and\, F_{nz}$. The simultaneous constraint method (Gardner 2001) is employed to solve Eq. 5.13. Figure 5.5 shows a diagram of full dynamic simulation of this system. There is no input system. The initial conditions (Eqs. 5.1 and 5.2) will determine its response. This matrix equation is embedded in an ODEs solution system of soccer ball motion. If accelerations $\ddot{x}$, $\ddot{y}$, $and\, \ddot{z}$ are integrated, then the velocities ($\dot{x}$, $\dot{y}$ and $\dot{z}$) and displacements (x, y, and z) will be computed. And the system will be available to solve the force equations and find aerodynamic drag $\vec{F}_d$, drag Magnus force $\vec{F}_m$ and contact force $\vec{F}_n$. In the next two sections, the work will be completed to simulate and analyze the ball shooting at a target.

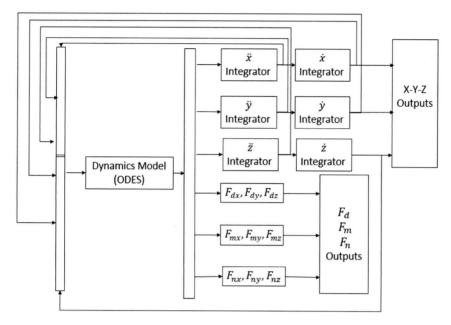

Fig. 5.5 Full dynamic simulation process of soccer ball free kick

5.3.2 Dynamic Simulation

Virtual prototype technology is applied for the dynamic simulation of soccer ball flight. The ball movement is simulated subject to all the underlying constraints to visualize and analyze the motions and forces. Figure 5.6a and b shows two cases of the superimposed display of the deployment history of the ball flight in a virtual field environment. The motions and forces are visualized by plotting successive ball positions on graphic displays, which include curving, bending, and spinning. The

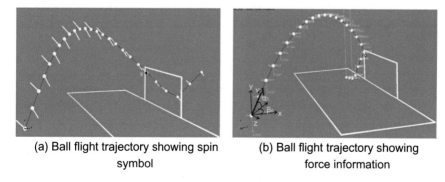

(a) Ball flight trajectory showing spin symbol	(b) Ball flight trajectory showing force information

Fig. 5.6 Dynamic simulation of a soccer ball motion with successive ball positions in a virtual field environment

dynamic force can be shown in x-y-z directions. The numerical results of the motion can be plotted and analyzed to examine the trajectory. The initial parameters can be changed for improving the dynamics performance of the ball flight. The trajectory can be optimized using parametric design, parametric simulation, and parametric analysis.

5.4 Simulation Results and Analysis

The actual processes of striking a soccer ball either with or without spin are simulated to understand how free kicks are performed. The dynamic model of the ball is tested to trace its displacements and velocities and to capture its dynamics performance.

Figure 5.7a and b shows the dynamic simulation of a soccer ball versus time 2.5 s. An un-massed bar is added to the ball as a symbol to show its spin (see Fig. 5.7b). The animation displays a curling trajectory with ball spinning. During the flight, Magnus effect makes the ball curve so that the flight trajectory exhibits a banana shape with a spatial curve.

Figure 5.8a and d reveals the variation of displacements of the ball in global coordinate $O(X, Y, Z)$ versus time 2.5 s. Figure 5.8a indicates that the ball's instantaneous position can be found in global coordinate $O(X, Y, Z)$. The displacements of the ball in three planes: (a) X-Y plane, (b) X-Z plane, and (c) Z-Y plane are plotted in Fig. 5.8b–d. The results show that the ball is initially at rest at the field. A player kicks the ball off its center with initial projectile angle $\theta_0 = 59.47°$ and initial orientation angle $\beta_0 = 35.73°$, causing it to fly with initial velocity $v_0 = 12.51$ m/s, and spin with initial angular velocity $\omega_0 = 31.42$ rad/s. When the ball reaches the goal at target B, its position $X = 10$ m, $Y = 2.04$ m, and $Z = -7$ m.

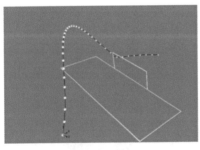 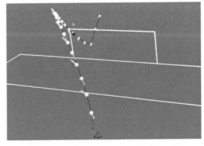

(a) Flight trajectory	(b) Flight trajectory, showing spin symbol

Fig. 5.7 Dynamic simulation of a soccer ball flight versus time 2.5 s with initial velocity $v_0 = 12.51$ m/s, initial projectile angle $\theta_0 = 59.47°$, initial orientation angle $\beta_0 = 35.73°$, and initial angular velocity $\omega_0 = 31.42$ rad/s

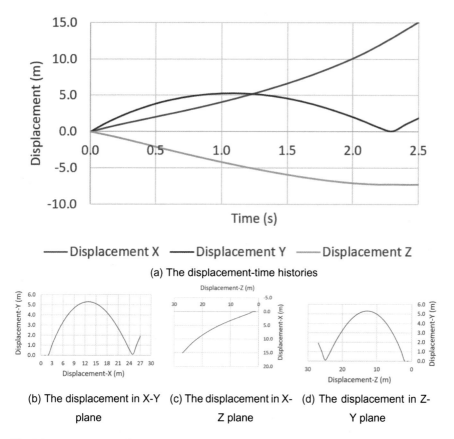

(a) The displacement-time histories

(b) The displacement in X-Y plane   (c) The displacement in X-Z plane   (d) The displacement in Z-Y plane

Fig. 5.8 The soccer ball displacements versus time 2.5

Figure 5.9 depicts the velocities of the ball versus time 2.5 s in global coordinate O(X, Y, Z). The magnitude of velocity can be measured from its three scalar components in x-y-z directions. For example, at time 2.25 s, the ball arrives at target B and its velocity = 13.7 m/s.

Figure 5.10 shows a ball is initially at rest at the field. A player kicks ball at its center with initial velocity $v_0 = 12.92$ m/s, initial projectile angle $\theta_0 = 58.48°$, and initial orientation angle $\beta_0 = 33.36°$. The target is set at point B (10, 2.04, −7) located on the goal at the left-top corner. The animation displaying trajectory reveals the ball instantaneous position in 3D space. The ball travels along a complex path. The flight trajectory shows the ball flies to the goal close to point B, but bounces off the goalpost, hits on the field, and then bounces up.

Comparison of two cases in Figs. 5.7 and 5.10 indicate that kicking ball off its center and at its center will lead the different traveling trajectories. In Fig. 5.10, the ball flies without spin, and then in Fig. 5.7, the ball flies with spin. Due to Magnus effect, when the ball flies, air flows over the ball. If the air flows faster around left side, it makes less pressure on the left side of the ball, because the faster the air

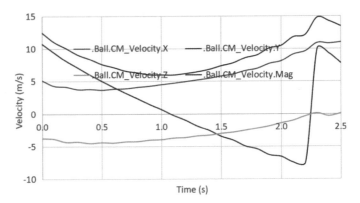

Fig. 5.9 The soccer ball velocities versus time 2.5 s

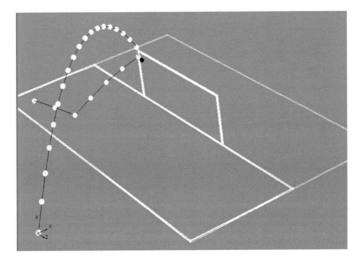

Fig. 5.10 Dynamic simulation of a soccer ball flight versus time 2.5 s with initial velocity $v_0 =$ 12.92 m/s, initial projectile angle $\theta_0 = 58.48°$, and initial orientation angle $\beta_0 = 33.36°$

flows, the lower the pressure of air becomes. On the right side of the ball, the air flows slower, as the spin is going directly against the flow of the air, so that there is much more pressure on the right side of the ball. The ball is pushed in the direction from high pressure to low pressure, making the ball curve.

Figure 5.11 shows dynamic simulation of a soccer ball flight versus time 2.5 s. The ball is launched at rest. A player kicks the ball off its center in the right side (s = 0.0237 m) with initial velocity $v_0 = 12.51$ m/s, initial projectile angle $\theta_0 = 59.47°$, initial orientation angle $\beta_0 = 35.73°$, and initial angular velocity $\omega_0 = 31.42$ rad/s. The target is set at point B (10, 2.04, − 7) located on the goal at the left-top corner. The ball traveling trajectory exhibits a standard banana shape. The flight trajectory shows the ball flies to target B, hits on the field, and then bounces into the net.

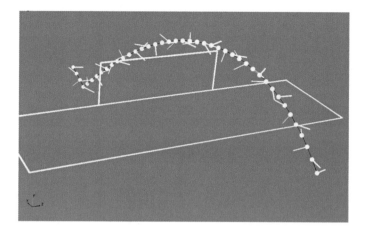

Fig. 5.11 Dynamic simulation of a soccer ball flight versus time 2.5 s with initial velocity $v_0 = 12.51$ m/s, initial projectile angle $\theta_0 = 59.47°$, initial orientation angle $\beta_0 = 35.73°$, and initial angular velocity $\omega_0 = 31.42$ rad/s

Figure 5.12 shows dynamic simulation of soccer ball flight versus time 2.5 s. The ball is launched at rest. A player kicks the ball off its center in the left side (s = 0.019 m) with initial velocity $v_0 = 14.48$ m/s, initial projectile angle $\theta_0 = 49.82°$, initial orientation angle $\beta_0 = 24.63°$, and initial angular velocity $\omega_0 = -31.42$ rad/s. The target point B (10, 2.04, −7) is set on the goal at the left-top corner. The ball traveling trajectory exhibits a standard banana shape. The ball flies to target B, hits on the field, and then bounces into the net.

Comparison of two cases in Figs. 5.11 and 5.12 can find the effects of spin on the flight trajectory. For kicking ball in the right side and the left side with the same

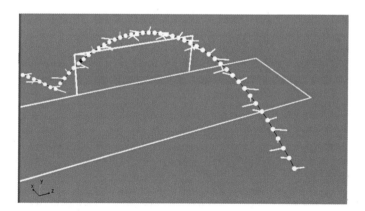

Fig. 5.12 Dynamic simulation of a soccer ball flight versus time 2.5 s with initial velocity $v_0 = 14.48$ m/s, initial projectile angle $\theta_0 = 49.82°$, initial orientation angle $\beta_0 = 24.63°$, and initial angular velocity $\omega_0 = -31.42$ rad/s

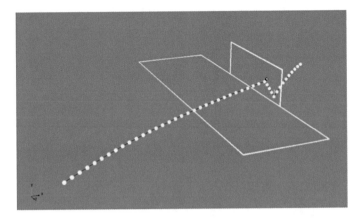

Fig. 5.13 Dynamic simulation of soccer ball flight versus time 2.2 s with initial velocity $v_0 = 9.83$ m/s, initial projectile angle $\theta_0 = 8.98°$, initial orientation angle $\beta_0 = 30.19°$, and initial angular velocity $\omega_0 = 31.42$ rad/s

angular velocity, it will require different velocities to reach a target. When kicking the ball in the right side, it will cause right spin on the ball with an angular velocity. When kicking the ball in the left side, it will cause left spin on the ball with the same angular velocity, but in the opposite direction. The Magnus forces are applied on the ball with the same magnitude, but in the opposite directions, making the ball curl in the opposite directions.

Figure 5.13 shows dynamic simulation of soccer ball flight versus time 2.2 s. A player kicks the ball off its center with initial velocity $v_0 = 9.83$ m/s, initial projectile angle $\theta_0 = 8.98°$, initial orientation angle $\beta_0 = 30.19°$, and initial angular velocity $\omega_0 = 31.42$ rad/s. When goalkeeper moves to the right side of the goal, the ball glanced off him, lands on the field, and bounds into the net. The flight path is changed two times.

Figure 5.14 shows dynamic simulation of soccer ball flight versus time 2.2 s. A player kicks the ball off its center with an initial velocity $v_0 = 9.94$ m/s, initial projectile angle $\theta_0 = 12.43°$, initial orientation angle $\beta_0 = 30.94°$, and initial angular velocity $\omega_0 = 31.42$ rad/s. The ball flight trajectory exhibits a similar shape with the one in Fig. 5.13. It shows the ball flies to the right side of the goal, glanced off the goal crossbar, lands on the field, and bounds into the net. The path of the ball flight is changed two times.

Comparison of two cases in Figs. 5.13 and 5.14 indicates that kicking ball with the same initial velocity, initial angular velocity, and initial orientation angle, but different initial project angle will affect the height of flight trajectory. The bigger the initial project angle is, the higher the ball flies. The initial project angle is a key to determine the flight trajectory and height and it should be defined correctly. The trajectory of soccer ball is sensitive to the initial conditions. To shoot at target, it requires to define a right velocity vector (magnitude and angles).

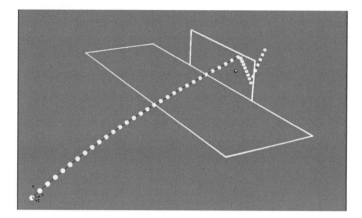

Fig. 5.14 Dynamic simulation of soccer ball flight versus time 2.2 s with initial velocity $v_0 = 9.94$ m/s, initial projectile angle $\theta_0 = 12.43°$, initial orientation angle $\beta_0 = 30.94°$, and initial angular velocity $\omega_0 = 31.42$ rad/s

The above results illustrate how closely a player must control the initial parameters of the kick to shoot at a target successfully. These cases illustrate that spin is an important determining factor in the trajectory of a rapidly moving ball. It is usually deliberately applied in the act of kicking, throwing or striking the ball when the player intends to modify the resulting flight (Bray and Kerwin 2003). The intention is to deceive an opponent by swerving a free kick in soccer. The Magnus effect causes the deflecting force due to the spin of a moving ball. The case studies in Chaps. 2–4 indicate that the flight trajectory of a moving but non-rotating ball is symmetrical about the line of flight due to the airflow separating at equivalent points around the ball's surface. The cases in this Chapter show that the trajectory of a flying and rotating ball is non-symmetrical due to the force caused by Magnus effect. The direction of the force is normal to the plane containing the velocity vector and the spin axis of the ball. The influence of spin on a soccer ball's flight at a corner kick will be discussed in next the chapter.

References

Asai T, Seo K, Kobayashi O, Sakashita R (2007) Fundamental aerodynamics of the soccer ball. Sports Eng 10:101–109

Asai T, Seo K (2013) Aerodynamic drag of modern soccer balls. Springerplus 2:171

Bush JWM (2013) The aerodynamics of the beautiful game. In: Clanet C (ed) Sports physics. Les Editions de l'Ecole Polytechnique, pp. 171–192

Bray K, Kerwin DG (2003) Modelling the flight of a soccer ball in a direct free kick. J Sport Sci 21:75–85

Goff JE, Carré MJ (2009) Trajectory analysis of a soccer ball. Am J Phys 77:1020–1027

Goff JE, Carré MJ (2010) Soccer ball lift coefficients via trajectory analysis. Eur J Phys 31:775–784

Gardner J (2001) Simulations of machines using Matlab and Simulink, 1st edn. University of California, Brooks

Li Y, Meng J, Li Q, (2020) Predicting soccer ball target through dynamic simulation. J. Eng. Res. Rep 12(4):6–18

Smith MR, Hilton DK, Van Sciver SW (1999) Observed drag crisis on a sphere in flowing He I and He II. Phys Fluids 11:751–753

Synge JL, Griffith BA (1959) Principles of mechanics. McGraw Hill, New York

Chapter 6
Modeling and Simulation of Soccer Ball Corner Kick

6.1 Overview

In this chapter, an innovative multi-body dynamics modeling of the soccer field-goal-ball-player has been proposed by integrating Newton's section law and Hooke's law. The model has mostly embedded the gravity, aerodynamic drag, drag moment, Magnus force, impulsive force, and contact force. The model allows randomly applying a kicking force or initial velocity on the ball in 3D space. The case studies of indirect and direct corner kick are focused on simulating reality and analyzing results. Virtual simulation provides a basic method in predicting the trajectory and examining dynamic performance of a soccer ball motion.

6.2 Initial Configuration

Figure 6.1 shows a dynamics model of a soccer ball to simulate a corner kick. The ball starts from the corner of field at point A, goes through the penalty area at point B, and reaches the goal at point C. The model includes multiplying solid bodies: field, goal, ball, and player. The components are made of different materials and are regarded as a multi-body system (Li et al. 2020). They are related by some restrictions. The field is fixed in the ground and provides a static platform for holding the goal, ball, and player. The field integrates the penalty area, which is the bigger box close to the goal, and the goal area, which is the small box inside the penalty area. The goal is structured by three bars, one crossbar, and two goalposts, and mounted on the field by the feet of two goalposts.

Table 6.1 summarizes the geometric parameters of field-goal-ball-player. They are (1) the soccer field: length 120 m and width 90 m; (2) the penalty area: length

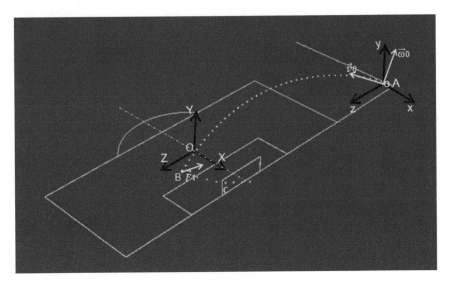

Fig. 6.1 Dynamics modeling of a soccer ball flight

Table 6.1 The geometric parameters of field-goal-ball-player

Geometric parameters										
Field		Penalty area		Goal area		Goal		Ball		Player
Length	Width	Length	Width	Length	Width	Width	Height	Diameter	Mass	Height
LF (m)	WF (m)	LP (m)	WP (m)	LA (m)	WA (m)	WG (m)	HG(m)	d (m)	m (kg)	HP (m)
120	90	40.32	16.5	28.32	5.5	7.32	2.44	0.2286	0.43	2.22

40.32 m and width 16.5 m; (3) the goal area: length 18.32 m and width 5.5 m; (4) the soccer goal: width 7.32 m and height 2.44 m; (5) the soccer ball: diameter d = 0.2286 m and mass m = 0.43 kg, and a player at point B: height 2.22 m.

A global Cartesian coordinate system O(X, Y, Z) is attached to the field at its center line and 10 m away from the goal, where X-Z plane is attached on the field with X-axis along the length direction, Z-axis along the width direction, and Y-axis is perpendicular to the field.

A local Cartesian coordinate system o(x, y, z) is attached to the soccer ball at its center of mass with x-y-z axis along X-Y-Z axis directions, respectively.

Figure 6.1 also shows the curling trajectory of the ball versus (vs) time. To curl the trajectory, the player kicks the ball off its center with initial force $\overrightarrow{F_0}$, causing it to fly with an initial velocity and spin with an initial angular velocity. Figure 6.2 indicates a free-body diagram of the ball with kicking force $\overrightarrow{F_0}$ applied at point P_0 in local coordinate o(x, y, z). The position and direction of force $\overrightarrow{F_0}$ depend on the radius of the ball r, projectile angle θ_0 and orientation angle β_0. To establish dynamics

Fig. 6.2 Free-body diagram
of the ball with a kicking
force and equivalent moment

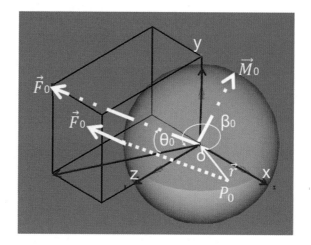

equations, force $\vec{F_0}$ is parallelly moved to the center of the ball o and an equivalent moment $\vec{M_0} = \vec{F_0} \times \vec{r}$ is generated and applied to point o. Its direction depends on $\vec{F_0} \times \vec{r}$.

If the foot and ball are in contact with time Δt, initial velocity $\vec{v_0}$ can be calculated by Eq. 6.1.

$$\vec{v_0} = \frac{\vec{F_0} \Delta t}{m} \tag{6.1}$$

Figure 6.3 indicates a diagram of the ball with initial velocity vector $\vec{v}_0$, initial angular velocity vector $\vec{\omega}_0$, initial projectile angle θ_0, and initial orientation angle β_0. Here, initial velocity vector $\vec{v}_0$ is written in three scalar components v_{x0}, v_{y0}, and v_{z0} in x-y-z directions in Eqs. 6.2a–c.

Fig. 6.3 Free-body diagram
of the ball with an initial
velocity and angular velocity

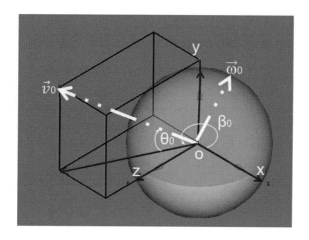

$$v_{x0} = v_0 \cos\theta_0 \cos\beta_0 \tag{6.2a}$$

$$v_{y0} = v_0 \sin\theta_0 \tag{6.2b}$$

$$v_{z0} = -v_0 \cos\theta_0 \sin\beta_0 \tag{6.2c}$$

Initial angular velocity $\vec{\omega}_0$ can be calculated by Eq. 6.3.

$$\vec{\omega}_0 = \frac{\vec{M}_0 \Delta t}{I} = \frac{\vec{v}_0 \times \vec{r}}{\frac{2}{3}r^2} \tag{6.3}$$

where I is the moment of inertial of the ball $I = \frac{2}{3}mr^2$, and $\vec{r}$ is the radius vector of the ball from point P_0 to point o. The direction of initial angular velocity $\vec{\omega}_0$ depends on $\vec{v}_0 \times \vec{r}$. Similarly, $\vec{\omega}_0$ is separated into three scalar components ω_{x0}, ω_{y0}, and ω_{z0} in x-y-z directions in Eqs. 6.4a–c.

$$\omega_{x0} = -\omega_0 \sin\theta_0 \cos\beta_0 \tag{6.4a}$$

$$\omega_{y0} = \omega_0 \cos\theta_0 \tag{6.4b}$$

$$\omega_{z0} = \omega_0 \sin\theta_0 \sin\beta_0 \tag{6.4c}$$

In this case study, the solution only focuses on finding initial velocity $\vec{v_0}$ and initial angular velocity $\vec{\omega}_0$ when given kicking force $\vec{F_0}$.

6.3 Dynamics Modeling

The dynamics model of the soccer ball is established under the assumptions: (1) the aerodynamic drag applied on the ball to be homogenous, and (2) neglecting the gyroscopic moment and acceleration of Coriolis. When a ball travels through the air, Fig. 6.4 shows a free-body diagram of the ball in an instantaneous position. The forces applied on the ball include the gravity $\vec{G}$, aerodynamic drag $\vec{F}_d$, drag moment $\vec{M}_d$, Magnus force $\vec{F}_m$, impulsive force $\vec{F}_t$, and contact force $\vec{F}_n$. Its dynamics equation is governed by Newton–Euler Equations (Synge and Griffith 1959).

$$\sum \vec{N} = m\vec{a} \tag{6.6a}$$

$$\sum \vec{M} = I\vec{\varepsilon} \tag{6.6b}$$

Fig. 6.4 Free-body diagram of a soccer ball in an instantaneous position

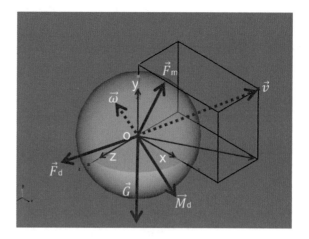

where $\sum \vec{N}$ is the vector summation of the general external forces; m is the mass of the ball; $\vec{a}$ is the acceleration vector of the ball; $\sum \vec{M}$ is the vector summation of the general external moments; I is the moment of inertia about the center of mass of the ball; $\vec{\varepsilon}$ is the angular acceleration vector of the ball. The motions and forces applied to the ball are established below.

Gravity: gravity $\vec{G}$ acts to the ball in -y-direction and is calculated by Eq. 6.7.

$$\vec{G} = -mg \tag{6.7}$$

where g is the gravitational acceleration and g = 9.8 m/s^2.

Aerodynamic Drag: the aerodynamic drag $\vec{F}_d$ acts to the ball in the direction opposite to the ball flying velocity $\vec{v}$ and is calculated by Eq. 6.8 (Bush 2013; Bray and Kerwin 2003; Asai et al. 2007, 2013; Goff and Carré 2009).

$$\vec{F}_d = -\frac{1}{2}\rho C_d A \vec{v} \lfloor \vec{v} \rfloor = -K_d \vec{v} |\vec{v}| \tag{6.8}$$

where ρ is the density of the air, $\rho = 1.205$ kg/m^3 for the environment temperature to be 20 °C; A is the cross-sectional area of the ball; and C_d is the drag coefficient. The drag coefficient depends on the boundary conditions, such as the surface roughness of a ball, as well as the laminar flow or turbulence of the atmospheric layer. Smith et al. (1999) suggested $C_d = 0.2$. If d denotes the diameter of the ball, then,

$$K_d = \frac{1}{8}\rho C_d \pi d^2 \tag{6.9}$$

Then, the aerodynamic drag $\vec{F}_d$ is broken down into three component forces, F_{dx}, F_{dy}, and F_{dz}, in x-y-z directions, respectively. Therefore, vector Eq. 6.8 can be written as three sets of scalar Eqs. 6.10a–c.

$$F_{dx} = -K_d \dot{x} |\dot{x}| \tag{6.10a}$$

$$F_{dy} = -K_d \dot{y} |\dot{y}| \tag{6.10b}$$

$$F_{dz} = -K_d \dot{z} |\dot{z}| \tag{6.10c}$$

Drag Moment: the drag moment $\vec{M}_d$ acts to the ball in the direction opposite to the angular velocity $\vec{\omega}$ and is calculated by Eq. 6.11.

$$\vec{M}_d = -\frac{1}{2} \rho C_{dm} A \vec{v}^2 \frac{\vec{\omega}}{\lfloor \vec{\omega} \rfloor} = -K_{dm} \vec{v}^2 \frac{\vec{\omega}}{\lfloor \vec{\omega} \rfloor} \tag{6.11}$$

where C_{dm} is the drag moment coefficient. Javorova and Ivanov (2018) suggested $C_{dm} = 0.05$. If coefficient K_{dm} is expressed as

$$K_{dm} = \frac{1}{8} \rho C_{dm} \pi d^2 \tag{6.12}$$

Then, the drag moment $\vec{M}_d$ is broken down into three components, M_{dx}, M_{dy}, and M_{dz}, in x-y-z directions, respectively. Therefore, vector Eq. 6.11 can be written as three sets of scalar Eqs. 6.13a–c.

$$M_{dx} = -K_{dm} \dot{x}^2 \frac{\omega_x}{|\omega_x|} \tag{6.13a}$$

$$M_{dy} = -K_{dm} \dot{y}^2 \frac{\omega_y}{|\omega_y|} \tag{6.13b}$$

$$M_{dx} = -K_{dm} \dot{z}^2 \frac{\omega_z}{|\omega_z|} \tag{6.13c}$$

Magnus Force: the Magnus force $\vec{F}_m$ acts to the ball in the direction of the $\vec{\omega} \times \vec{v}$ (Gupta and Panigrahi 2013) It is perpendicular to both the angular velocity $\vec{\omega}$ and velocity $\vec{v}$, and is calculated by Eq. 6.14.

$$\vec{F}_m = \frac{1}{2} \rho C_m A \vec{v}^2 \frac{\vec{\omega} \times \vec{v}}{\lfloor \vec{\omega} \times \vec{v} \rfloor} = K_m \vec{v}^2 \frac{\vec{\omega} \times \vec{v}}{\lfloor \vec{\omega} \times \vec{v} \rfloor} \tag{6.14}$$

where C_m is the Magnus coefficient. Goff and Carre (2010) suggested $C_m = 0.2$. If coefficient K_m is expressed as

$$K_m = \frac{1}{8}\rho C_m \pi d^2 \tag{6.15}$$

Then, the Magnus force $\overrightarrow{F}_m$ is separated into three components, F_{mx}, F_{my}, and F_{mz}, in x-y-z directions, respectively. Therefore, vector Eq. 6.14 can be written as sets of scalar Eqs. 6.16a–c.

$$F_{mx} = K_m \dot{z}^2 \frac{\omega_y v_z}{\lfloor \omega_y v_z \rfloor} - K_m \dot{y}^2 \frac{\omega_z v_y}{\lfloor \omega_y v_y \rfloor} \tag{6.16a}$$

$$F_{my} = K_m \dot{x}^2 \frac{\omega_z v_x}{\lfloor \omega_z v_x \rfloor} - K_m \dot{z}^2 \frac{\omega_x v_z}{\lfloor \omega_x v_z \rfloor} \tag{6.16b}$$

$$F_{mz} = K_m \dot{y}^2 \frac{\omega_x v_y}{\lfloor \omega_x v_y \rfloor} - K_m \dot{x}^2 \frac{\omega_y v_x}{\lfloor \omega_y v_x \rfloor} \tag{6.16c}$$

Impulsive Force: impulsive force $\overrightarrow{F}_t$ is applied to the ball by a player in penalty area. Its direction depends on the projectile angle θ_t and orientation angle β_t. Referring to Fig. 6.1, when the ball arrives to location B, a player heads or kicks the ball off its center, causing it to reach the goal at point C. Figure 6.5 indicates a free-body diagram of the ball with an impulsive force $\overrightarrow{F}_t$ applied at point P_t in local coordinate o(x, y, z). In order to establish dynamics equations, force $\overrightarrow{F}_t$ is parallelly moved to the center of the ball o and an equivalent moment $\overrightarrow{M}_t$ is generated and applied to point o. $\overrightarrow{F}_t$ is expressed as three scalar components F_{tx}, F_{ty}, and F_{tz} in x-y-z directions by Eqs. 6.17a–c.

$$F_{tx} = F_t \cos \theta_t \cos \beta_t \tag{6.17a}$$

Fig. 6.5 Free-body diagram of the ball with impulsive force and equivalent moment

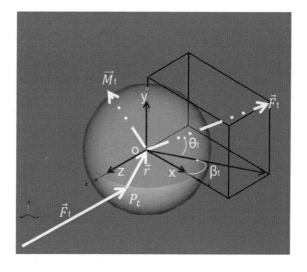

$$F_{ty} = F_t \sin \theta_t \qquad (6.17b)$$

$$F_{tz} = -F_t \cos \theta_t \sin \beta_t \qquad (6.17c)$$

The moment $\overrightarrow{M}_t$ can be calculated by Eq. 6.18.

$$\vec{M}_t = \vec{F}_t \times \vec{r} \qquad (6.18)$$

where $\overrightarrow{r}$ is the radius vector from point P_t to point o. The direction of moment $\overrightarrow{M}_t$ depends on $\overrightarrow{F}_t \times \overrightarrow{r}$. Similarly, it is separated into three scalar components M_{tx}, M_{ty}, and M_{tz} in x-y-z directions in Eqs. 6.19a–c.

$$M_{tx} = -F_t r \sin\theta_t \cos \beta_t \qquad (6.19a)$$

$$M_{ty} = F_t r \cos\theta_t \qquad (6.19b)$$

$$M_{tz} = F_t r \sin\theta_t \sin \beta_t \qquad (6.19c)$$

Contact Force: when a ball contacts with a body such as a player, field, goal post (see Fig. 6.6), a contact force $\overrightarrow{F}_n$ is generated between the ball and the body. Its direction is normal to the contact surfaces at contact point. Figure 6.6 shows a model

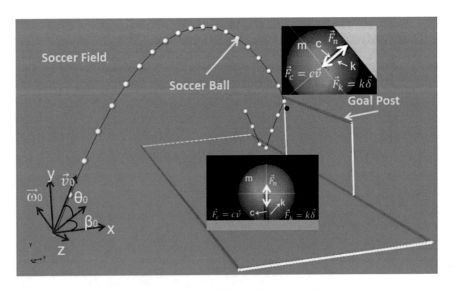

Fig. 6.6 Contact force between the ball and field or the ball and goal post

of the ball-body contact with the mass-dumping-stiffness properties. The equation of the contact force is derived by employing Newtown's Second Law and Hooke's law.

$$\vec{F}_n = \vec{F}_c + \vec{F}_k = c\vec{v} + k\vec{\delta} \tag{6.20}$$

where $\vec{F}_c$ is the damping force vector, $\vec{F}_k$ is the elastic force vector, c is the contact damping coefficient (c = 3.8×10^5 N/m), $\vec{v}$ is the velocity vector, k is the contact stiffness (k = 97.2 N.s/m), $\vec{\delta}$ is the displacement vector. Note the contact damping coefficient and contact stiffness cannot find in material property handbooks. The data given here are calculated results. The contact force $\vec{F}_n$ is separated into three scalar components F_{nx}, F_{ny}, and F_{nz} in x-y-z directions in Eqs. 6.21a–c.

$$F_{nx} = c\dot{x} + kx \tag{6.21a}$$

$$F_{ny} = c\dot{y} + ky \tag{6.21b}$$

$$F_{nz} = c\dot{z} + kz \tag{6.21c}$$

So far, Eq. 6.6a can be written as

$$\sum \vec{N} = m\vec{a} = \vec{G} + \vec{F}_d + \vec{F}_m + n\vec{F}_t + e\vec{F}_n \tag{6.22}$$

where, (1) n = 0, for no contact occurring between the ball and head and the impulsive force is zero; (2) n = 1, for the contact occurring between the ball and head and the impulsive force is larger than zero; (3) e = 0, when $\vec{\delta}$ >= d/2, there is no penetration between the ball and body and the contact force is zero; and (4) e = 1, when $\vec{\delta}$ < d/2, there is the penetration between the ball and body and the contact force is larger than zero.

Equation 6.22 is separated into differential Equations in x-y-z directions.

$$m\ddot{x} = -K_d\dot{x}|\dot{x}| + K_m\dot{z}^2 \frac{\omega_y v_z}{\lfloor \omega_y v_z \rfloor} - K_m\dot{y}^2 \frac{\omega_z v_y}{\lfloor \omega_y v_y \rfloor} + nF_t\cos\theta_t\cos\beta_t + e(c\dot{x} + kx) \tag{6.23a}$$

$$m\ddot{y} = -mg - K_d\dot{y}|\dot{y}| + K_m\dot{x}^2 \frac{\omega_z v_x}{\lfloor \omega_z v_x \rfloor} - K_m\dot{z}^2 \frac{\omega_x v_z}{\lfloor \omega_x v_z \rfloor} + nF_t\sin\theta_t + e(c\dot{y} + ky) \tag{6.23b}$$

$$m\ddot{z} = -K_d\dot{z}|\dot{z}| + K_m\dot{y}^2 \frac{\omega_x v_y}{\lfloor \omega_x v_y \rfloor} - K_m\dot{x}^2 \frac{\omega_y v_x}{\lfloor \omega_y v_x \rfloor} - nF_t\cos\theta_t\sin\beta_t + e(c\dot{z} + kz) \tag{6.23c}$$

Also, Eq. 6.6b can be written as

$$\sum \vec{M} = I\vec{\varepsilon} = \vec{M}_d + n\vec{M}_t \tag{6.24}$$

Equation 6.24 is broken into differential Equations in x-y-z directions.

$$I\dot{\omega}_x = -K_{dm}\dot{x}^2 \frac{\omega_x}{|\omega_x|} - nF_t r \sin\theta_t \cos\beta_t \tag{6.25a}$$

$$I\dot{\omega}_y = -K_{dm}\dot{y}^2 \frac{\omega_y}{|\omega_y|} + nF_t r \cos\theta_t \tag{6.25b}$$

$$I\dot{\omega}_z = -K_{dm}\dot{z}^2 \frac{\omega_z}{|\omega_z|} + nF_t r \sin\theta_t \sin\beta_t \tag{6.25c}$$

Assembling Eqs. 6.23a–c and 6.25a–c in matrix form, the general dynamics model is represented as Eq. 6.26.

$$\begin{bmatrix} m & 0 & 0 & 0 & 0 & 0 \\ 0 & m & 0 & 0 & 0 & 0 \\ 0 & 0 & m & 0 & 0 & 0 \\ 0 & 0 & 0 & 1 & 0 & 0 \\ 0 & 0 & 0 & 0 & 1 & 0 \\ 0 & 0 & 0 & 0 & 0 & 1 \end{bmatrix} \begin{bmatrix} \ddot{x} \\ \ddot{y} \\ \ddot{z} \\ \dot{\omega}_x \\ \dot{\omega}_y \\ \dot{\omega}_z \end{bmatrix}$$

$$= \begin{bmatrix} -K_d\dot{x}|\dot{x}| + K_m\dot{z}^2 \frac{\omega_y\, v_z}{[\omega_y\, v_z]} - K_m\dot{y}^2 \frac{\omega_z\, v_y}{[\omega_y\, v_y]} + nF_t \cos\theta_t \cos\beta_t + e(c\dot{x} + kx) \\ -mg - K_d\dot{y}|\dot{y}| + K_m\dot{x}^2 \frac{\omega_z\, v_x}{[\omega_z\, v_x]} - K_m\dot{z}^2 \frac{\omega_x\, v_z}{[\omega_x\, v_z]} + nF_t \sin\theta_t + e(c\dot{y} + ky) \\ -K_d\dot{z}|\dot{z}| + K_m\dot{y}^2 \frac{\omega_x\, v_y}{[\omega_x\, v_y]} - K_m\dot{x}^2 \frac{\omega_y\, v_x}{[\omega_y\, v_x]} - nF_t \cos\theta_t \sin\beta_t + e(c\dot{z} + kz) \\ -K_{dm}\dot{x}^2 \frac{\omega_x}{|\omega_x|} - nF_t r \sin\theta_t \cos\beta_t \\ -K_{dm}\dot{y}^2 \frac{\omega_y}{|\omega_y|} + nF_t r \cos\theta_t \\ -K_{dm}\dot{z}^2 \frac{\omega_z}{|\omega_z|} + nF_t r \sin\theta_t \sin\beta_t \end{bmatrix}$$

$$\tag{6.26}$$

This matrix forms a kinematics and dynamics system of six equations and six unknowns, $\ddot{x}$, $\ddot{y}$, $\ddot{z}$, $\dot{\omega}_x$, $\dot{\omega}_y$, and $\dot{\omega}_z$. The simultaneous constraint method (Gardner 2001) is employed to solve Eq. 6.26. Figure 6.7 shows the dynamic simulation diagram of this system. There is no input system. The initial conditions (Eqs. 6.2–6.4) will determine its response. This matrix equation is embedded in an ODEs solution system of soccer ball motion. If accelerations $\ddot{x}$, $\ddot{y}$, $and\ddot{z}$ and angular accelerations $\dot{\omega}_x$, $\dot{\omega}_y$, $and\, \dot{\omega}_z$ are integrated, then the velocities $(\dot{x},\ \dot{y},$ and $\dot{z})$ and displacements (x, y, and z), angular velocities $(\omega_x, \omega_y, and\, \omega_z)$ and angular displacements, $(\theta_x, \theta_y\ and\ \theta_z)$ will be available to compute the matrix and right side of the equation.

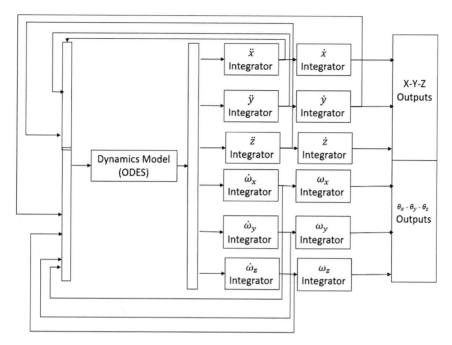

Fig. 6.7 Flowchart of dynamic simulation process of soccer ball conner kick

Furthermore, combining Eqs. 6.10a–c, 6.13a–c, 6.16a–c, and 6.21a–c to form an assembling matrix Eq. 6.27. the dynamic dynamics system can be extended to solve the force equations and find aerodynamic drag $\overrightarrow{F}_d$, drag moment $\overrightarrow{M}_d$, Magnus force $\overrightarrow{F}_m$ and contact force $\overrightarrow{F}_n$. In the next section, the work will implement the dynamic simulation and analysis of the ball shooting at target.

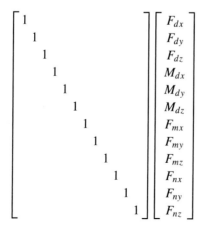

$$\begin{bmatrix} 1 \\ & 1 \\ & & 1 \\ & & & 1 \\ & & & & 1 \\ & & & & & 1 \\ & & & & & & 1 \\ & & & & & & & 1 \\ & & & & & & & & 1 \\ & & & & & & & & & 1 \\ & & & & & & & & & & 1 \\ & & & & & & & & & & & 1 \end{bmatrix} \begin{bmatrix} F_{dx} \\ F_{dy} \\ F_{dz} \\ M_{dx} \\ M_{dy} \\ M_{dz} \\ F_{mx} \\ F_{my} \\ F_{mz} \\ F_{nx} \\ F_{ny} \\ F_{nz} \end{bmatrix}$$

$$= \begin{bmatrix} -K_d \dot{x} |\dot{x}| \\ -K_d \dot{y} |\dot{y}| \\ -K_d \dot{z} |\dot{z}| \\ -K_{dm} \dot{x}^2 \frac{\omega_x}{|\omega_x|} \\ -K_{dm} \dot{y}^2 \frac{\omega_y}{|\omega_y|} \\ -K_{dm} \dot{z}^2 \frac{\omega_z}{|\omega_z|} \\ K_m \dot{z}^2 \frac{\omega_y v_z}{\lfloor \omega_y v_z \rfloor} - K_m \dot{y}^2 \frac{\omega_z v_y}{\lfloor \omega_y v_y \rfloor} \\ K_m \dot{x}^2 \frac{\omega_z v_x}{\lfloor \omega_z v_x \rfloor} - K_m \dot{z}^2 \frac{\omega_x v_z}{\lfloor \omega_x v_z \rfloor} \\ K_m \dot{y}^2 \frac{\omega_x v_y}{\lfloor \omega_x v_y \rfloor} - K_m \dot{x}^2 \frac{\omega_y v_x}{\lfloor \omega_y v_x \rfloor} \\ c\dot{x} + kx \\ c\dot{y} + ky \\ c\dot{z} + kz \end{bmatrix} \qquad (6.27)$$

6.4   Dynamic Simulation and Results Analysis

The actual processes of soccer ball shooting at goals are simulated to understand how corner kicks are performed. The dynamic models of the ball are evaluated to trace the displacements and velocities and to capture the dynamics performance.

6.4.1   Indirect Corner Kick

Usually, the classic implementation of corner kick is an indirect score, which consists of two steps. At the corner of field, a striker kicks the ball with spin to cause it to fly along a curling trajectory. When the ball reaches the penalty area, his teammate touches the ball and shoots at goal.

Figure 6.8a–g shows the dynamic simulation of a soccer ball versus time 2.5 s. The animation display of curling trajectory provides four views: (a) isometric view, (b) top view, (c) side view, and (d) front view. The displacement of the ball is displayed in a global coordinate O(X, Y, Z) with three planes: (e) X-Z plane, (f) Z-Y plane, and (g) X-Y plane.

Figure 6.8a shows the ball is initially at rest at the field corner A, where X = 9 m, Y = 0 m, and Z = −34 m (see Fig. 6.8e–f). A player kicks the ball at point P_0 (0.1074, 0, 0.039) with a kick force $\vec{F}_0 = 604.58$ N, projectile angle $\theta_0 = 30°$, and orientation angle $\beta_0 = 250°$, causing it to fly and spin. If the foot and ball are in contact with time $\Delta t = 0.02$ s, the ball flies with initial velocity $v_0 = 28.12$ m/s and spin with initial angular velocity $\omega_0 = 31.42$ rad/s.

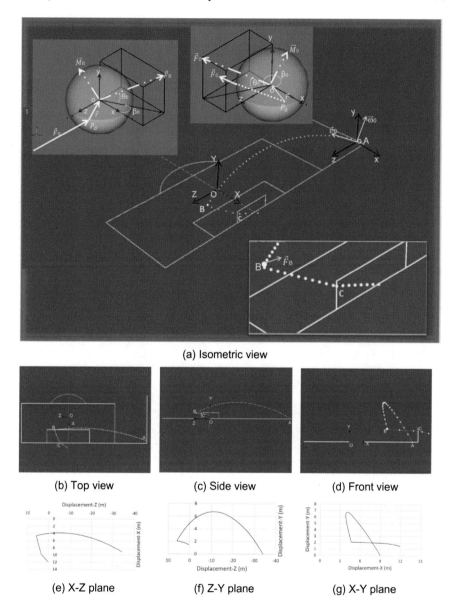

(a) Isometric view

(b) Top view   (c) Side view   (d) Front view

(e) X-Z plane   (f) Z-Y plane   (g) X-Y plane

Fig. 6.8 Dynamic simulation of a soccer ball flight versus time 2.5 s

A player is at penalty area B, where $X = 4.61$ m, $Y = 2.22$ m, and $Z = 5.72$ m as given in Fig. 6.8e–f. During the flight, Magnus effect makes the ball curve. From field corner A to penalty area B, the ball flight trajectory exhibits a banana shape with a spatial curve (see Fig. 6.8b–d).

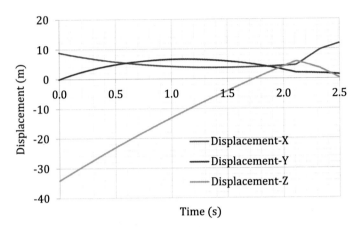

Fig. 6.9 The soccer ball displacements versus time 2.5 s

At position B, A player heads the ball off its center with force $\overrightarrow{F_B}$, causing it to change the flight trajectory toward to the goal. Impulsive force $\overrightarrow{F_B} = 758$ N applied to the ball at point $P_B(0.085, 0, 0.076)$, where r = 0.1143 m and projectile angle θ_B = 14.05° and orientation angle β_B = 48.30°.

When the ball reaches to the goal at point C (X = 10 m, Y = 1.87 m, and Z = 3.35 m as shown in Fig. 6.8e–f), the ball contacts with the goal post (see Fig. 6.8a). The flight trajectory is changed again, and the ball glanced off the goal post into the net.

Figure 6.9 reveals the variation of displacements of the ball in global coordinate O(X, Y, Z) versus time 2.5 s. The ball's instantaneous position is plotted in global coordinate O(X, Y, Z). The ball is initially located at field corner A(9, 0, −34) at time 0 s. At time 2.13 s, the ball moves to location B(4.61, 2.22, 5.72) and the tendencies of displacements are changed due to a player touching the ball. At time 2.34 s, the ball arrives to the goal at point C(10, 1.87, 3.35) and the tendencies of displacements are changed again because of the ball contacts the goalpost.

Figure 6.10 depicts the velocities of the ball versus time 2.5 s in global coordinate O(X, Y, Z). Figure 6.11 indicates a free-body diagram of the ball with instantaneous velocities in local coordinate o(x, y, z). The magnitude of velocity v_t can be measured from its three scalar components v_{tx}, v_{ty}, and v_{tz} in x-y-z directions. The direction of velocity v_t is defined by its projectile angle θ_t and orientation angle β_t. In Fig. 6.10, it notes that the ball is launched with an initial velocity $v_0 = 28.12$ m/s ($v_{0x} = -8.33$ m/s, $v_{0y} = 14.06$ m/s, $v_{0z} = 22.88$ m/s) with $\theta_{v0} = 30°$ and $\beta_{v0} = 250°$ (see Fig. 6.12a). At time 2.13 s, the ball moves to location B and its velocity $v_B = 27.3$ m/s ($v_{Bx} = 25.23$ m/s, $v_{By} = -0.61$ m/s, $v_{Bz} = -10.54$ m/s) with $\theta_{vB} = -1.28°$, and $\beta_{vB} = 22.67°$ (see Fig. 6.12b). At time 2.34 s, the ball reaches the goal at point C and its velocity $v_C = 22.6$ m/s ($v_{Cx} = 12.4$ m/s, $v_{Cy} = -1.47$ m/s, $v_{0z} = -18.88$ m/s) with $\theta_{vC} = 14.05°$ and $\beta_{vC} = 48.30°$ (see Fig. 6.12c).

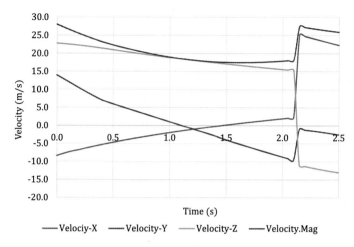

Fig. 6.10 The soccer ball velocities versus time 2.5 s

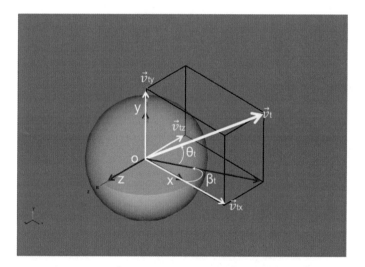

Fig. 6.11 Free-body diagram of the ball with instantaneous velocities

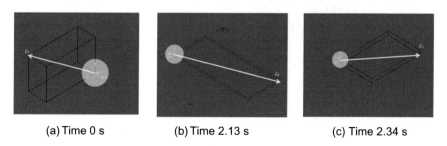

(a) Time 0 s   (b) Time 2.13 s   (c) Time 2.34 s

Fig. 6.12 Free-body diagram of the ball with velocity vectors at selected times

6.4.2 Direct Corner Kick

Also, a corner kick can cause a direct score, which just takes one step to shoot at goal. Figures 4.11, 4.12 and 4.13 gives the dynamic simulation of a soccer ball versus time in three styles of corner kick.

Figure 6.13 shows the dynamic simulation of a soccer ball flight versus time 0.8 s. The corner striker launches the ball with velocity $v_0 = 60$ m/s, projectile angle $\theta_0 = 12°$, orientation angle $\beta_0 = 190°$, and angular velocity $\omega_0 = 31.42$ rad/s. Figure 6.13a shows that the ball travels along a banana trajectory and goes through goal. Figure 6.13b plots the velocity–time history of the ball flight. In this case, the ball spins with a constant angular velocity $\omega_0 = 31.42$ rad/s.

Figure 6.14 shows the dynamic simulation of a soccer ball flight versus time 2.4 s without drag moment. The corner striker launches the ball with velocity $v_0 = 28$ m/s, projectile angle $\theta_0 = 30°$, orientation angle $\beta_0 = 258.75°$, and angular velocity $\omega_0 = 31.42$ rad/s. Figure 6.14a shows that the ball travels along a spatial curve, goes exactly to the target, hits on the field, and then bounces into the net. Figure 6.14b plots the velocity–time history of the ball flight. In this case, the ball spins with a constant angular velocity $\omega_0 = 31.42$ rad/s.

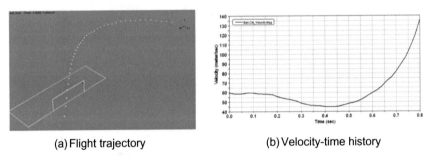

(a) Flight trajectory   (b) Velocity-time history

Fig. 6.13 Dynamic simulation of a soccer ball versus time 0.8 s

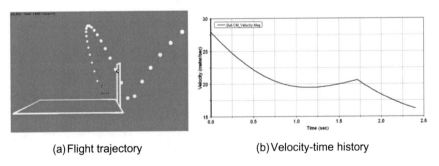

(a) Flight trajectory   (b) Velocity-time history

Fig. 6.14 Dynamic simulation of a soccer ball versus time 2.4 s without drag moment

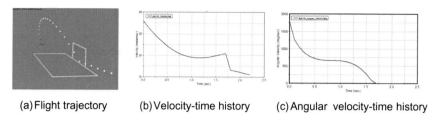

(a) Flight trajectory   (b) Velocity-time history   (c) Angular velocity-time history

Fig. 6.15 Dynamic simulation of a soccer ball versus time 2.2 s with drag moment

Figure 6.15 shows the dynamic simulation of a soccer ball flight versus time 2.2 s with drag moment. The corner striker launches the ball with velocity $v_0 = 28$ m/s, projectile angle $\theta_0 = 30°$, orientation angle $\beta_0 = 258.75°$, and angular velocity $\omega_0 = 31.42$ rad/s. Figure 6.15a displays that the ball travels along a spatial trajectory, goes through the goal, hits on the field, and then bounces into the net. Figure 6.15b gives the velocity–time history of the ball flight. In this case, the ball spins with variable angular velocity as shown in Fig. 6.15c. The angular velocity decreases from 31.42 to 12.22 rad/s (1800–700 deg/s) with the time from 0 to 0.5 s, stays on 12.22 rad/s (700 deg/s) from time 0.5–1.2 s, and then decreases to 0 rad/s (0 deg/s) from time 1.2 to 2.2 s.

The effort of Magnus effect on the trajectory can be found in Figs. 4.11, 4.12 and 4.13. When the corner striker launches the ball with a velocity and angular velocity, causing it to fly and spin. During the ball traveling, air moves over the ball and Magnus effect makes the ball curve to form a curling trajectory. When the ball travels with spin, the spin can cause some effects on the ball flight. During spinning, the air goes faster over the ball with pressure under-neath. This will cause the ball to rise and travel farther. As shown in Figs. 4.11, 4.12 and 4.13, after going through the goal, the ball hits the field and then bounces off the field. It goes almost straight, as there is less air resistance when the ball is traveling fast. When the ball slows down, it starts to curl as the Magnus effect occurs. The slower the ball moves, the more it curls.

The effect of drag moment on the trajectory can be seen by comparing Figs. 6.14 and 6.15. The drag moment affects the ball that flies and spins through the air. When the corner striker launches the ball with the same velocity, projectile angle, orientation angle, and angular velocity, the ball travels with different trajectories as shown in Figs. 6.14a and 6.15a, respectively. In the former, the ball flies with a constant angular velocity 31.42 rad/s. In the latter, the ball flies with variable angular velocity as shown in Fig. 6.15c. When the ball travels, air moves over the ball and resists it spinning so that the angular velocity decreases with time. Often engineering dynamics problems used in teaching ignore the drag moment, but it is particularly important for understanding the motion of fast-moving-spinning ball. The drag moment depends on the density of the air, the area of the ball, the angular velocity it is spinning, and the drag moment coefficient.

References

Asai T, Seo K, Kobayashi O, Sakashita R (2007) Fundamental aerodynamics of the soccer ball. Sports Eng 10:101–109

Asai T, Seo K (2013) Aerodynamic drag of modern soccer balls. Springerplus 2:171

Bush JWM (2013) The aerodynamics of the beautiful game. In: Clanet C (ed) Sports physics. Les Editions de l'Ecole Polytechnique, pp 171–192

Bray K, Kerwin DG (2003) Modelling the flight of a soccer ball in a direct free kick. J Sport Sci 21:75–85

Goff JE, Carré MJ (2009) Trajectory analysis of a soccer ball. Am J Phys 77:1020–1027

Goff JE, Carré MJ (2010) Soccer ball lift coefficients via trajectory analysis. Eur J Phys 31:775–784

Gardner J (2001) Simulations of machines using Matlab and Simulink, 1st edn. University of California, Brooks

Gupta G, Panigrahi PK (2013) Curve kick aerodynamics of a soccer ball. In: Proceeding of the fortieth national conference on fluid mechanics and fluid power, Himachal Pradesh

Javorova and Ivanov, 2018. Javorova J, Ivanov A (2018) Study of soccer ball flight trajectory. MATEC Web Conf 145:01002

Li Y, Meng J, Li Q (2020) Predicting soccer ball target through dynamic simulation. J. Eng. Res. Rep 12(4):6–18

Smith MR, Hilton DK, Van Sciver SW (1999) Observed drag crisis on a sphere in flowing He I and He II. Phys Fluids 11:751–753

Synge JL, Griffith BA (1959) Principles of mechanics. McGraw Hill, New York

Chapter 7
Contributions and Conclusions

In this book, dynamics modeling, optimization design and virtual simulation of soccer ball have been introduced for efficient sports analysis. The studies combine dynamics, mathematics, and sports engineering from a multi-disciplinary design optimization toward soccer ball spatial kinematics and dynamics simulation. A foundation is built for the full dynamic simulation of soccer ball motion. Chaps. 3–6 discuss five subjects: model validation, projectile motion, trajectory optimization, free kick, and corner kick. Each of these is independent and highlights different aspects of soccer ball kinematics and dynamics. The book clearly remarks problems, objectives, methodology, and main results. The basic modeling equations are supported with strong references and their physical meanings are clearly explained.

The publications related to the previous research are introduced in Chap. 1. What was done in previous works has been described and the main results of each of them have been stressed. To help the readers to properly understand how the present work compares to the subject in this book. The exact contributions of the works in this book are discussed and compared to recently published work. Some important conclusions are summarized, and future works are suggested.

7.1 Dynamics Modeling

The general modeling process involves the initial configuration, dynamics modeling, dynamic simulation, and result analysis. The initial conditions can be a kick force or an initial velocity, which are calculated employing the principle of linear impulse and momentum. The multi-body dynamics model of the soccer field-goal-ball-player has been established by integrating Newton's second law and Hooke's law. The model involves the gravity, aerodynamic drag, drag moment, Magnus force, impulsive force, and contact force.

Y. Li, *Motion Analysis of Soccer Ball*,
SpringerBriefs in Applied Sciences and Technology,
https://doi.org/10.1007/978-981-16-8652-8_7

In Chaps. 2 and 3, a spatial two-body dynamics model of a soccer field-ball has been developed. The model included gravity, aerodynamic drag, and contact force. The equation of the contact force between the ball and field is integrated into the model. In the previous works, dynamics model mostly was created as a free-body dynamics model and the ball was the only body to take into account. The research focused on the studies about aerodynamics characteristics without contact force item.

In Chap. 2, the method of numerical model validation is introduced by using a simple example. The model is validated by the comparison of the simulation result and a theoretical calculation result (Hroncová and Grieš 2014). A good agreement has been obtained between the two models. The most theoretical studies do not include model validation. Actually, it is very important to verify if the model is reliable to provide the exact simulation results.

In Chaps. 4 and 5, a spatial kinematics and dynamics model is developed for the study of soccer free kick. The two-body dynamics model is extended to the three-body dynamics model of a soccer field-goal-ball, and the Magnus force is added in. The model allows randomly applying a kicking force or an initial velocity on the ball in 3D space. At this point, similar work has been published by Zhu et al. (2017). They have designed a training system to simulate the free kick of a soccer ball. The trajectory can be predicted for the given initial parameters. However, the work has focused on the kinematics simulation rather than the dynamics simulation. Therefore, the dynamic contact force of field-ball doesn't include in their model.

In Chap. 6, a spatial four-body dynamics model of a soccer field-goal-ball-player is developed for the simulation of direct or indirect conner kick. The model involves the gravity, aerodynamic drag, drag moment, Magnus force, impulsive force, and contact force. The motion of the soccer ball is expressed as ODEs. Javorova and Ivanov (2018) have created a soccer ball model to simulate the direct conner kick. The aerodynamic forces and moments applied to the ball are considered. But the kicking force and contact force of field-ball doesn't include in their model. In their study, the ODEs which describe the motion of the soccer ball are solved numerically by MATLAB-Simulink software. The solution method lacks to be illustrated in detail. Why is such method used? Undergraduate students may be interested in the solution methodology. In this book, using some typical examples, the numerical solution of the ODEs is described step by step.

In general, classic mechanics has been implemented into the soccer ball modeling. The innovative multi-body dynamics modeling of the soccer field-goal-ball-player has been proposed by integrating Newton's section law and Hooke's law. The model has mostly embedded the gravity, aerodynamic drag, drag moment, Magnus force, impulsive force and contact force.

7.2 Optimization Design

An optimization method has been proposed to predict a soccer ball target. The equations of motion include the items of aerodynamic drag and Magnus force and gravity. To optimize the traveling trajectory, the design objective has been modeled as the mathematical equations with the design variables and design constraints. This model is able to capture the simultaneous distance between a ball and target to guide a direct optimization. In such a manner, the optimal trajectory can be identified for players of distinct ability.

In Chap. 4, an example is given to demonstrate the application of the optimization design method on the soccer ball shooting at goal. The results indicate how closely the player must control the initial parameters of the kick to achieve a successful outcome. From the modeling through simulation to optimization, it is concluded that the ball motions can be displayed visually, measured numerically, and optimized parametrically. The result shows that the most optimal combination of the design parameters goes to initial velocity $v_0 = 12.844$ m/s, initial projectile angle $\theta_0 = 57.705°$, and initial orientation angle $\beta_0 = 34.575°$ for achieving the ball to reach the target. Therefore, the method is useful in monitoring the trajectory and improving the initial parameters.

The optimization design is an interesting subject in sports analysis. The optimization problem usually is about solving an optimal trajectory. This study integrates parameterization technology into the optimization design can exactly capture the expected flight trajectory of a soccer ball. Some studies only focused on theoretical modeling. Some studies just armed to improve the original trajectory. Anyway, it is very important to provide a general procedure to describe the flowchart of optimization operation for the advanced sports analysis.

7.3 Dynamic Simulation

The dynamic simulation has focused on the case studies: soccer ball projectile motion, trajectory optimization, free kick, and corner kick. In Chap. 3, an example is given to indicate the kinematics and dynamics simulation of a soccer ball projectile motion in a virtual environment. In kinematics simulation & analysis, the influences of initial velocity on the time-related projectile height and range have been analyzed and discussed. In dynamics simulation & analysis, the effects of aerodynamic drag on the flight trajectory over a time interval have been investigated. The impulsive force applied on the soccer ball has been discovered for producing an expected initial velocity. All works form a complete procedure to illustrate the application of dynamic simulation on soccer ball projectile motion analysis.

In Chaps. 4–6, the parameterization is used for the modeling and simulation of soccer ball flight. This function is required for the full dynamic simulation of the soccer ball shooting target. It allows randomly positioning a ball as well as applying a kick force or an initial velocity and angular velocity on the ball in 3D space. In recent works (Zhu et al. 2017; Javorova and Ivanov 2018), the simulation systems have been developed to simulate the soccer free kick. The trajectory can be predicted for the given initial parameters. However, their works have focused on the kinematics simulation and didn't mention the parameterization problem.

The 3D virtual prototypes are built to animate the 3D motions. The motion of the ball can be visualized by plotting successive ball positions on graphic displays. The instantaneous positions of the ball are clearly displayed. The ball's curving, bending, and spinning postures can be captured in the instantaneous position. The instantaneous force applied to the ball is visualized and plotted with the simulating time, which is useful to analyze the dynamic performance. Therefore, the virtual prototype technology is very useful to visualize soccer ball motion and dynamic force distribution in the 3D space. There were a lot of studies about flight trajectories. The results have output as the diagrams of displacement–time, but it may be hard to imagine the position of the ball in the 3D space. It is hard to find the distribution of dynamic forces.

In generation, this book provides valuable guidance for the dynamics modeling, model validation, multi-disciplinary design optimization, and virtual prototype simulation of soccer ball and further improvement. Some typical examples are given to indicate the practical applications. The kinematics and dynamics simulation of a soccer ball are addressed and how to analyze the results are described. One of the purposes of this book is about the sports analysis. Therefore, the detailed result analysis and discussion are emphasized. The most previous studies have focused on the development of a simulation system but lacked case studies to explain how to use the system, and how to analyze the simulation results.

7.4 Future Works

This book is significant to guide to model, simulate, and analyze the soccer ball motion for optimizing trajectory and accurate position. Three aspects require to be improved through further investigation. They are: (1) model improvement, such as extending the current equations of motion by adding more factors, which can be an acceleration of Coriolis and gyroscopic moment; (2) validation extension, such as validating the dynamics model by comparing the experimental results with the simulated results; and (3) case-study diversification, such as performing the case studies with unsteady-state airflow for the further illustrations of kinematics and dynamics simulation.

References

Hroncová D, Grieš M (2014) Trajectories of projectiles launched at different elevation angles and modify design variable in MSC Adams/view. Appl Mech Mater 611:198–207. ISSN: 1662-7482

Javorova and Ivanov, 2018. Javorova J, Ivanov A (2018) Study of soccer ball flight trajectory. MATEC Web Conf 145:01002

Zhu ZQ, Chen B, Qiu SH, Wang RX (2017) Simulation and modeling of free kicks in football game and analysis on assisted training. In: Asian simulation 2017, part I, CCIS751. IEEE, pp 413–427

Printed in the United States
by Baker & Taylor Publisher Services